REMOTE SENSING FOR TROPICAL ECOSYSTEM MANAGEMENT

Proceedings of the Sixth Regional Seminar on Earth Observation for Tropical Ecosystem Management
Ho Chi Minh City, Viet Nam, 3-7 November 1997

Jointly organized by
the National Space Development Agency of Japan,
the National Centre for Science and Technology of Viet Nam
and the Economic and Social Commission for Asia and the Pacific

UNITED NATIONS
New York, 1997

Front cover: JERS-1 SAR and optical data colour composite, Ho Chi Minh City. Courtesy of Ministry of International Trade and Industry and National Space Development Agency of Japan.

ST/ESCAP/1914

UNITED NATIONS PUBLICATION

Sales No. E.99.II.F.32

ISBN: 92-1-119901-8

TECHNICAL EDITOR'S NOTE

In editing this report, every effort has been made to retain the original contents as presented by the authors. Editorial adjustments have sometimes been made in contents and/or illustrations which do not seriously affect the original form of the presentations.

Preface

The Sixth Regional Seminar on Earth Observation for Tropical Ecosystem Management was held at Ho Chi Minh City, Viet Nam, from 3 to 7 November 1997. This annual seminar has been organized by the National Space Development Agency (NASDA) and the Remote Sensing Technology Center (RESTEC) of Japan in cooperation with ESCAP since 1992. It has been held in six different countries of the Asian and Pacific region: Thailand in 1992, Malaysia in 1993, Indonesia in 1994, the Philippines in 1995, Fiji in 1996 and Viet Nam in 1997. The ESCAP secretariat is extremely pleased to have collaborated with NASDA and RESTEC in the organization of this useful activity and would like to take this opportunity to extend its gratitude to those agencies on behalf of the members and associate members of ESCAP.

Like the previous seminars, the sixth seminar was designed in two distinct parts. The first part aimed to create awareness among senior-level managers and planners concerned with natural resources and environmental management about the usefulness of remote sensing and geographic information system (GIS) techniques in mapping, monitoring and managing the land and its natural resources in tropical environments. Several well-known experts from NASDA, the Asian Institute of Technology, the European Space Agency and the University of Hannover joined other experts from the region and contributed by providing examples of the use of optical and microwave remote sensing as well as GIS for solving practical problems that are typically encountered in tropical environments. The second part of the seminar consisted of hands-on-training in digital image-processing and GIS techniques for junior-level scientists and technologists. Its purpose was to either complement their existing skills or introduce them to these important technological tools. This hands-on-training component was designed to provide practical knowledge and skills to practitioners from various government agencies working in natural resources and environment management.

This seminar is only one of the many activities organized each year by ESCAP, in cooperation with other agencies and institutions, which aim to assist in national capacity-building and are undertaken under the Regional Space Applications Programme for Sustainable Development, which was launched by the Ministerial Conference on Space Applications for Development in Asia and the Pacific, held at Beijing in 1994.

The ESCAP secretariat believes that activities such as this one significantly contribute to widening the base of users of remote sensing and GIS, spatial information technologies that can no longer be ignored if natural resources and the environment are going to be managed effectively and efficiently.

CONTENTS

Page

Preface iii

Abbreviations vi

PART ONE. REPORT OF THE SEMINAR

I. Report of the Sixth Regional Seminar on Earth Observation for Tropical Ecosystem Management 3

PART TWO. PAPERS: FUTURE PROSPECTS FOR REMOTE SENSING

II. Earth observation in the multimedia age 15
Shunji Murai

III. Future prospects for mapping from space 21
Gottfried Konecny

IV. Japanese Earth observation project 36
Takashi Moriyama

V. Remote sensing and GIS in Viet Nam 38
Tran Manh Tuan

PART THREE. PAPERS: TROPICAL ECOSYSTEMS

VI. Depletion of mangrove forest and its causes in Thailand 45
Darasri Dowreang

VII. Forest mapping in Viet Nam 47
Nguyen Manh Cuong

VIII. Reforestation following an ecological approach 64
Sirin Kawla-ierd

IX. Degradation of the Upper Mahaweli catchment and its impact 67
H. Manthrithilake

X. Management of ecosystems in Myanmar 71
Ohn Gyaw

PART FOUR. PAPERS: MICROWAVE REMOTE SENSING

XI. European Remote Sensing satellite, six years in orbit: the mission and selected applications 77
Robert Schumann

XII. Signal processing of JERS-1 SAR data 81
Makoto Ono

XIII. Application of microwave images to wetland and coastal zone monitoring 88
Supapis Polngam

ANNEX

List of participants 107

ABBREVIATIONS

ADEOS	Advanced Earth Observing Satellite (Japan)
AG/IGOS	Analysis Group/International Global Observing Strategy
ALOS	Advanced Land Observation System (Japan)
ANSI	American National Standards Institute
AutoCAD	Automated Computer-assisted Design
AVHRR	Advanced Very High Resolution Radiometer
AVNIR	Advanced Visible and Near-Infrared Radiometer
BPI	Bits per inch
CAL/VAL	Calibration/validation
CCT	Computer-compatible tape
CD-ROM	Compact disc-read-only memory
CEOS	Committee on Earth Observation Satellites
DBH	Diameter-base-height
DEM	Digital elevation model
EOSAT	Earth Observation Satellite Company
ERS	European Remote Sensing
ESA	European Space Agency
FAO	Food and Agriculture Organization of the United Nations
FFT-IFFT	Fast fourier transform – Inverted fast Fourier transform
GIS	Geographic information system
GMS	Geostationary Meteorological Satellite (Japan)
GOES	Geostationary Operational Environmental Satellite (United States of America)
GPS	Global Positioning System
HH	Horizontal emit and receive
HV	Horizontal emit and vertical receive
ILWIS	Integrated Land and Water Information System
IR	Infra-red
IRS	Indian Remote Sensing satellite
ISO/TC	International Organization for Standardization Technical Committee
JERS	Japan Earth Resources Satellite
KVR	High-resolution camera (Russian Federation)
LISS	Linear Imaging Self-scanning Sensor
MESSR	Multispectral Electronic Self-scanning Radiometer
MOMS	Modular Opto-electronic Multispectral Scanner
MOS	Marine Observation Satellite (Japan)
MSS	Multispectral scanner
NASA	National Aeronautics and Space Administration (United States)
NASDA	National Space Development Agency (Japan)
NCST	National Centre for Science and Technology (Viet Nam)
NOAA	National Oceanic and Atmospheric Administration (United States)
OPS	Optical sensor
PAN	Panchromatic sensor
SAR	Synthetic aperture radar
SPANS	Spatial Analysis System (GIS software)
SPOT	Système pour l'observation de la Terre (France)
SWIR	Short wave infra-red
TM	Thematic Mapper
UNDP	United Nations Development Programme
UTM	Universal Transverse Mercator projection
VH	Vertical emit and horizontal receive
VV	Vertical emit and receive
WGISS	Working Group on Information Systems and Services (CEOS)

PART ONE
REPORT OF THE SEMINAR

I. REPORT OF THE SIXTH REGIONAL SEMINAR ON EARTH OBSERVATION FOR TROPICAL ECOSYSTEM MANAGEMENT

ORGANIZATION

The Sixth Regional Seminar on Earth Observation for Tropical Ecosystem Management was held at Ho Chi Minh City, Viet Nam, from 3 to 7 November 1997. The Seminar was jointly organized by the National Space Development Agency of Japan (NASDA), the United Nations Economic and Social Commission for Asia and the Pacific (ESCAP) and the National Centre for Science and Technology of Viet Nam (NCST), Government of Viet Nam. The Remote Sensing Technology Center (RESTEC) of Japan and the Asian Center for Research on Remote Sensing of the Asian Institute of Technology (AIT) provided technical support, and the Government of Japan, through the NASDA cooperation programme, provided financial support for the Seminar.

OBJECTIVES

The major objectives of the Seminar were (a) to exchange scientific and technical information on Earth observation technology and its applications to sustainable ecosystem management and (b) to promote integration of remote sensing with development planning. The Seminar also aimed at providing hands-on training on the utilization of personal computer-based remote sensing and geographic information system (GIS) facilities for tropical ecosystem management in the developing countries.

PROGRAMME

The Seminar was organized in two parts:

(a) A three-day seminar (3-5 November 1997) for high-level planners on the applications of remote sensing and GIS for sustaining terrestrial ecosystem management, including a field trip to study a tropical ecosystem;

(b) Two days of hands-on training (6-7 November 1997) in remote sensing and GIS applications to tropical ecosystem management for working-level specialists from various agencies of the host country.

ATTENDANCE

The Seminar was attended by 60 participants and observers from 12 countries and 5 organizations: Bangladesh, Cambodia, China, Lao People's Democratic Republic, Malaysia, Myanmar, Philippines, Sri Lanka, Thailand, Viet Nam, Germany, Japan, AIT, European Space Agency (ESA), NASDA, RESTEC and ESCAP.

SESSIONS AND ACTIVITIES

Opening of the Seminar

The Seminar was opened by Nguyen Van Hieu, President of the National Centre for Science and Technology of Viet Nam. In this opening address, Professor Nguyen expressed his appreciation to all co-organizers for their support of the Seminar. Since ESCAP had launched its programmes on remote sensing and on space technology applications, Viet Nam had benefited significantly by participating in the various activities organized by ESCAP. He assured the members that the National Centre for Science and Technology of Viet Nam would continue its cooperation with

ESCAP and other interested institutions in promoting space applications, particularly remote sensing and GIS in Viet Nam, through closer regional cooperation.

In welcoming the participants to Ho Chi Minh City, Tran Thanh Long, Vice-President, People's Committee of Ho Chi Minh City, stated that the countries of the region were in the stages of socio-economic development and that the applications of space technology in the fields of forestry, agriculture and the study of natural disasters and many other fields of the economy had brought impressive results. The applications of remote sensing and GIS for tropical ecosystem management and environment protection were essential for sustainable development in Viet Nam. He expressed appreciation to the organizers for giving Ho Chi Minh City the opportunity to play host for that important Seminar.

Representing the co-organizers, Sohsuke Gotoh, Special Assistant to the President of NASDA, mentioned that certain progress had been made since the first regional seminar held in Thailand and that NASDA was happy to have played an important role in promoting the region in close cooperation with ESCAP and its member countries. He informed the Seminar that NASDA had started a pilot project on the operational use of satellite data in Thailand and that NASDA intended to extend the experience and results to other parts of the region step by step. He expressed satisfaction with the ESCAP secretariat for close cooperation in the past five years. He assured the participants that NASDA would strengthen its efforts in promoting operational uses of satellite data in the region.

A message from the Executive Secretary of ESCAP was presented by the Chief of the Space Technology Applications Section of ESCAP. It was stated that the Regional Seminar on Earth Observation was only one of the many activities organized each year by ESCAP in cooperation with other agencies and institutions which aimed to assist in national capacity-building through the Regional Space Applications Programme for Sustainable Development (RESAP), which had been launched by the Ministerial Conference on Space Applications for Development in Asia and the Pacific, held at Beijing in 1994. The Executive Secretary stressed that such activities, which significantly contributed to widening the base of users, should continue to be prioritized and supported, if the region wished to manage natural resources and the environment effectively and efficiently. The Executive Secretary expressed his gratitude to all parties, particularly NASDA, RESTEC and AIT, for their cooperation and support they had given in the organization of the Seminar and he also assured the Seminar of the continued cooperation and support of ESCAP for that annual event.

Following the formal opening session, Gottfried Konecny, University of Hannover, Germany, delivered a keynote speech entitled "Future prospects for mapping from space". He presented the major issues confronted by the world's communities in pursuing the objectives of sustainable development, emphasizing that population expansion and environmental degradation and natural disasters were the major problems threatening development. He argued that to support sustainable development planning, information on natural resources and the environment was essential and that remote sensing and GIS technology had become important and indispensable tools in addressing the information needs of developing countries. He analysed the demands for basic data sets, the current capability in remote sensing and GIS, and the future assistance that the new space technology could offer in meeting the needs. He stated that the world would become more interdependent in the new century and therefore international and regional cooperation in space applications should be strengthened.

Session 1: Future prospects of Earth observation

Shunji Murai, Chairman, Space Technology Applications and Research Program (STAR) of AIT, gave a presentation entitled "Earth observation in the multimedia age". He argued that the Earth observation technology was entering the third generation in conjunction with multimedia technology. Space technology applications, including not only space-borne remote sensing but also the Global Positioning System (GPS) and satellite communication, were now playing an important role in monitoring the Earth's environment, with support of geographic information systems (GIS). Nowadays continuous measurement of a position with a few millimetre accuracy was possible with GPS, while real-time positioning with 20-50 centimetre accuracy was available on board vehicles, boats, and airplanes. Within a year, very-high-resolution satellite imagery with one metre ground resolution would become commercially available. Those data would be transmitted to anywhere in the world through the Internet or satellite communication. On the other hand, in accordance with the advancement of digital image processing, such as animation and virtual reality, the results obtained from monitoring of the Earth's environment could be represented in more sensible forms that were more understandable.

Information on recent Japanese Earth satellite projects was presented by Takashi Moriyama of NASDA. The paper described the role of tropical ecosystems in Earth's environment and the role of satellite observation. An example of a global rainforest map using JERS-1 data, which had been developed by NASDA, was presented. NASDA would make efforts to continue its contribution to environment management by providing data to users through joint research and pilot projects. The paper also presented the near future plans of NASDA in Earth observation, especially those concerning high spatial resolution data acquisition, such as images in visible, near-infra-red and short wave infra-red (SWIR). The paper also argued that since there were many initiatives to develop and launch commercial or research satellites over time, it had become important to establish a user community on satellite use.

Tran Manh Tuan of NCST of Viet Nam gave a presentation entitled "Remote sensing and GIS in Viet Nam". He gave a brief overview about the main institutions in Viet Nam that were involved in remote sensing and GIS and their projects and results. Exploiting the Internet's remote sensing and GIS information, applying new techniques in remote sensing and GIS, establishing the national GIS for natural resource management, and environment monitoring on the network were the main directions for remote sensing and GIS in the future in Viet Nam.

Session 2: Tropical ecosystems

In her presentation entitled "Depletion of mangrove forest and its causes in Thailand", Darasri Dowreang of the National Research Council of Thailand (NRCT) discussed major problems related to mangrove forest depletion in Thailand. Mangrove forests were known to be an important ecosystem ecologically and economically. They were a source of wood and charcoal and also served as a breeding ground and nursery of terrestrial and aquatic fauna and therefore had high potential for the fishery industry, including fish and shrimp farming. Landsat TM images had been used to monitor changes in mangrove forests and it was found that mangrove area had decreased from 3,678 sq km in 1961 to 1,676 sq km in 1996, i.e. more than 50 per cent in 35 years. Shrimp farming was found to be one of the major causes, among other activities that included orienting and forest encroachment for settlements. In order to prevent the forest from being further degraded and depleted, new measures for mangrove forest management would be established under a five-year project for 1999-2003. Such measures included building awareness and concern; protection and privatization through local participation; rehabilitation and reforestation; and conducting research in areas such as biodiversity and reutilization of shrimp farms.

In his presentation entitled "Forest mapping in Viet Nam", Nguyen Manh Cuong, Forestry Institute of Planning and Investigation (FIPI) of Viet Nam, provided information on activities that had been carried out by the Ministry of Agriculture and Rural Development of Viet Nam and its long process in development. Since 1980 the remote sensing method had become the main method for forest mapping. The first forest map was established at country level in Viet Nam in 1980. After 16 years, many forest maps at different scales had been produced based on remote sensing data. The maps had been used as important information sources in many international and national development projects. The role of forest mapping based on remote sensing and GIS had become a more and more important and effective tool in Viet Nam.

In her paper entitled “Reforestation by ecological approach”, Sirin Kawla-ierd of the Royal Palace, Thailand, presented a case of forest restoration based on phytosociological surveys, which differed from conventional reforestation that had been attempted in Thailand since 1991. Phytosociological surveys, the first priority of the restoration system, were carried out to understand the forest ecology of natural and secondary forests. The phytosociological surveys had been used to restore and manage degraded forest areas. The six-year results showed that the growth performance (basal diameter and total height) of the secondary, late successional and climax species indicated their high potential in the restoration of degraded areas.

Herath Manthrithilake presented a paper entitled “Degradation of Upper Mahaweli catchment (UMC) and its impact”. The catchment area was an important area for the country in terms of water, energy and food production. Degradation of that area was a concern for everybody in the country. The paper highlighted the methods applied to develop an “indicative land degradation map” for UMC utilizing the GIS-based information collated in the Mahaweli Authority of Sri Lanka. It was hoped to extend the exercise to other parts of the country using remotely sensed data.

Session 3: Microwave remote sensing

Robert Schumann, representative of the European Space Agency, presented a paper entitled “European Remote Sensing satellite, six years in orbit: the mission and selected applications”. The paper provided a brief overview of the European Remote Sensing satellite (ERS) missions and emphasized continuity of the various payload elements into the Envisat and EUMETSAT Meteorological Operational satellite series (METOP) programmes. Various applications of one of the main ERS instruments, the synthetic aperture radar (SAR), were also reviewed with emphasis on sites in South-East Asia. Those examples included applications in agriculture, forestry and pollution, and the potential of techniques such as multitemporal combination and interferometry were discussed. The presentation ended with mention of a current announcement of opportunity (AO) for the exploitation of ERS data, which had a submission date of 1 February 1998, and advance notice of a similar AO for Envisat, due for release that month, along with the URL from which full texts of both AOs could be downloaded: <http://esa-ao.org>.

Presenting a paper entitled “Signal processing of JERS-1 SAR data”, Makoto Ono, of RESTEC, Japan, introduced a PC-based SAR processor, which runs all functions on a PC with occupation of a small disk area yet with enough speed to process JERS-1 SAR data. The programme was open domain and could be downloaded free of charge from the RESTEC Web site.

Emphasizing uses of microwave remote sensing, Supapis Polngam, from the National Research Council of Thailand, presented her study entitled “Application of microwave images to wetland and coastal zone monitoring”. The study showed that a mangrove swamp forest could be clearly distinguished from brackish swamp forest based on multi-sensor microwave images of ERS-1 and Radarsat. JERS-1 had diminished ability to recognize those two classes. However, JERS-1, with long wavelength L-band HH, was more sensitive that C-band VV and HH to standing tree classes, which had sharp boundaries. HH polarization of C band and L band gave good response for shrimp farms and floating weeds. Wave patterns, currents and shallow sea bathymetry provided strong radar signals from C-band VV polarization. Data fusion of multitemporal Radarsat data could reasonably support shrimp farm monitoring and rice growing stages.

Panel discussion: How to promote operational use of satellite data

A panel discussion on how to promote operational use of satellite data was held during the Seminar. The panel was chaired by Shunji Murai, Chairman of the Programme Committee of the Seminar. The five panelists who contributed to the subject were Pham Van Cu of NCST, Takashi Moriyama of NASDA, He Changchui of ESCAP, Robert Schumann of AIT and Darasri Dowreang of NRCT. Major topics covered during the panel discussion included data policy, pricing and services, technical support and consultancy services, education and training, institutional building, and communication and coordination. Gottfried Konecny, Nazmul Hoque, Eriberto Argete and Herath Manthrithilake also contributed to the panel discussion.

The panel recognized that human resource development was essential for wider and more efficient use of remote sensing data for sustainable development. The relationship between education, training and research was discussed and it was emphasized that those three aspects of technology development should be maintained in balance. In that connection, it was suggested that the Education and Training Network on remote sensing should be strengthened and that the centres of excellence in the region should be supported to promote education and training, particularly in advanced training courses.

The panel also recognized the important role of regional networking and coordination in promoting satellite data applications. Human networking at both national and regional levels was considered useful for enhancing communications between users, planners and decision makers. The role of national organizations and local authorities in coordinating remote sensing activities at their respective levels was indispensable. At the regional level, the role of ESCAP in promoting regional coordination and cooperation was recognized.

Possible solutions for major problems in data policy, data accessibility and data services were discussed. It was recognized that there was a need to establish a policy framework by the concerned government agencies to promote data utilization. The major dimensions of the framework would include the provision of data free of charge for research work and operational uses by non-commercial users; regional data nodes for wider distribution of data; and ramped price schemes for projects with commercial potential. The concept of EAT (easy access, affordability and timeliness) in data distribution was advocated and considered important in promoting wider use of remote sensing data.

The panel, while recognizing the importance of government policy in data investment and in supporting operational uses of remote sensing data, also believed that it was important to encourage the involvement of the private sector in developing the remote sensing data service industry. The panel held the view that, because of the economic downturn in the region, there would be increasing pressure on the concerned government agencies that would have to create a sustaining mechanism for remote sensing data distribution and data services. A quicker development of private industry in remote sensing services was envisaged by the panel.

Electronic networking of remote sensing data was considered important for exchange of information and sharing of remote sensing data for environment and natural resource management and disaster monitoring. The role of the Internet and homepages in promoting remote sensing applications was highlighted. The panel emphasized the need for developing national and international standards on remote sensing data, guidelines for sharing of data, and technical protocols for database interoperability. ESCAP, NASDA and other organizations such as the Committee on Earth Observation Satellites (CEOS) were invited to cooperate on those matters.

The panel stressed the need to improve communication between remote sensing technologists and users. Communication between remote sensing communities and government planners and policy makers was essential for operationalization of remote sensing. Efforts should be made to persuade policy makers to invest in space applications through awareness-creation activities such as high-level seminars, newspapers, television programmes and even Internet technology. Development of concise manuals and guidebooks for planners and decision makers was considered important and useful. Using good and effective remote sensing-derived information for emergency uses such as forest fire and flood monitoring was particularly effective in helping decision makers become more supportive of remote sensing. The panel was of the opinion that there should be a continued effort on that matter.

Demonstrations

A demonstration of Japanese satellite images was made by Shoji Takeuchi of RESTEC. The demonstration covered images from three satellites that were launched by NASDA, namely JERS-1, MOS and ADEOS. The demonstration used an image-processing software package called ASEAN (Advanced System for Environmental Analysis with remote sensing data), which was developed on the Windows 95 platform. The images covered mainly some areas in Thailand and Viet Nam. The participants were informed that NASDA and RESTEC had jointly developed and produced satellite data on CD-ROM, providing a satellite remote sensing image data set that could be easily accessible to the end users. That approach had facilitated the utilization of the data from Japanese Earth observation satellites. Through that demonstration, the participants obtained first-hand information on uses of satellite data in wide application areas such as land-use classification, forestry inventory, coastal zone monitoring (including mangrove mapping), urban planning, agriculture and ecosystem management. The potentiality of combinations of various satellite data, including the fusion of SAR data with optical images and the integration of remote sensing and GIS, was also presented.

The ASEAN software was developed by Nguyen Dinh Duong, Institute of Geography, NCST of Viet Nam, in 1992 under financial support from NASDA and technical cooperation with RESTEC. The system was further upgraded and an improved version, WinASEAN, was developed based on Microsoft Windows 3.1. WinASEAN could cover most basic image analysis functions, which were divided into pre-processing, image display, multispectral classification, post-classification, geometric correction, change analysis, bird's eye view and image overlay. WinASEAN was menu-driven by Microsoft Windows with a simple user-friendly interface that provided easy-to-use features for beginners. WinASEAN was released through the annual Regional Remote Sensing Seminar on Tropical Ecosystem Management as public domain software. WinASEAN had been further upgraded on Windows 95 with an additional GIS module. The software package of WinASEAN, along with 20-plus scenes of Japanese satellite images, had been realized on CD-ROM for public domain use.

Field trip

A technical tour to a national reserve area in Binh Chau was organized. The purpose of the field trip was to enable the participants to have a better appreciation of tropical ecosystems. It also provided an opportunity for better undestanding of how remote sensing and GIS technology were used to support tropical ecosystem management. The participants were guided in verifying ground truth with satellite images of the national reserve area.

Hands-on training

The hands-on training was conducted on 6 and 7 November 1997. The purpose of the training was to enhance national capacity in using PC-based image-processing and GIS techniques for handling remote sensing and GIS data. The training was undertaken by 12 local participants from nine government organizations, including the environment agency, agricultural department, geological department and provincial organizations.

The training covered a brief introduction to a PC-based image-processing system such as WinASEAN and utilization of WinASEAN for handling a CD-ROM data set, which was prepared by RESTEC using mainly Japanese satellite data in combination with data from several other state-of-the-art satellites. Some basics in remote sensing image processing, such as geometric correction, SAR data analysis and multispectral data classification were also introduced through hands-on practice. The integrated analysis of remote sensing and GIS data was demonstrated and practice sessions were arranged.

A CD-ROM containing the upgraded version of WinASEAN and the satellite data set was distributed free of charge to all training participants for their further self-learning and practice.

The participants felt that the training had been highly intensive and very useful and they thanked the organizers for providing the training opportunity.

CONCLUSIONS AND RECOMMENDATIONS

Conclusions

It was widely recognized that for sustainable development planning, accurate and up-to-date information on natural resources and the environment had become indispensable. It was considered that space technology, such as remote sensing and GIS, had become an important tool for addressing information needs in sustainable development planning.

Remote sensing and GIS technology had developed rapidly in recent years, and in the near future new technology would offer many more competitive functions such as high resolution and hyperspectral capability, frequent coverage and quick data distribution, as well as information networking for operational use.

Monitoring the Earth's environment by remote sensing would be rapidly commercialized because the technology had become operational and the private sector had become intensely involved in spatial information technology. A number of commercial satellite programmes would soon be available and there was a need for user communities to keep abreast of the new trends.

Multimedia technology would provide an opportunity for users to access information about the latest status of the Earth's environment and to predict changes. More high-value-added information would be produced by integrating remote sensing and GIS with visualization based on sensibility.

There was an increasing recognition of the need to establish and enhance regional infrastructure for spatial information networking through communication satellites and other communication technologies. Networking and emerging multimedia technologies would enable the region to benefit significantly from spatial information use in sustainable development.

There was a need for the development of policy guidelines and data standardization to facilitate acquisition and distribution of satellite data through networks.

Accompanied by rapid developments in satellite remote sensing sensor technology (such as high-resolution, hyperspectral imaging spectrometers, laser imaging scanners, and interferometer and SAR technology), computer processing software with artificial intelligence and multimedia technology would be further developed for real-time information extraction and production for operational use of remote sensing data.

For sustainable management of tropical ecosystems, it had become necessary for remote sensing and GIS specialists to adopt a multidisciplinary approach and integrate themeselves with local communities in implementing development projects.

To prepare the region for the operational uses and the challenge of the commercialization of Earth observation technology, the strengthening of human resource development and the enhancing of national capacity-building had become of paramount importance. Regional cooperation on user development through training, workshops, seminars, study and research, as well as pilot projects, continued to be an area of high priority in the region.

It was recognized that the joint efforts of NASDA and ESCAP in promoting regional cooperation through co-organizing the annual Regional Seminar on Earth Observation for Tropical Ecosystem Management, with technical support from RESTEC and AIT, had borne fruit and that the impact on national capacity-building in participanting countries had been tangible. Such cooperation needed to be continued and further strengthened.

Recommendations

Realizing the complexity of ecosystem management, it was recommended that the multidisciplinary approach and intersectoral cooperation should be emphasized as a solution to remedy the feeble situation of operational use of remote sensing data in ecosystem management.

Recognizing that education and training on tropical ecosystem management using remote sensing and GIS were essential for environmentally sound and sustainable development in the region, the Regional Seminar on Earth Observation for Tropical Ecosystem Management should be continued by NASDA in cooperation with ESCAP.

Acknowledging the reiterated offer of the Government of Bangladesh to host the Regional Seminar, it was further recommended that the next Seminar should be held in Bangladesh, with the host facility provided by the Space Research and Remote Sensing Organization (SPARRSO).

Recognizing the strong need for technical support and consultancy in research projects of developing countries in Asia and the Pacific, technical and funding support should be explored by potential donors such as NASDA to assist research activities in developing countries.

In that connection, the centres of excellence such as the Asian Center for Research on Remote Sensing (ACRoRS), which had been newly founded by the Asian Association on Remote Sensing (AARS) and AIT, should be supported to conduct training of researchers from the region.

In view of the usefulness of satellite remote sensing data for sustainable development of natural resources and the environment in Asia and the Pacific, space agencies such as NASDA should be encouraged to adopt an open data policy to promote satellite data use. The users, particularly the non-commercial users, should be given friendly access to data or other user service facilities at a nominal charge.

Recognizing that there was increasing demand for and strong interest in exchanging experiences and sharing information on Earth observation for ecosystem management, a manual/guidebook for ecosystem management using remote sensing and GIS should be developed and published, based on the information provided in that series of seminars. Interested donors, including NASDA, should be invited to provide support for that activity.

Acknowledging that there was increasing demand for advanced training courses in remote sensing and GIS for sustainable development, more funds should be allocated for conducting training in GIS applications, SAR interferometry, GPS surveys, stereo-matching and computer-assisted cartography.

Appreciating that mapping from space was becoming more and more important, and that it provided opportunities for integrating remote sensing and GIS for the construction of GIS databases, increased attention should be paid to cooperation in education, training and research in computer mapping technology.

Since information sharing was seen to be essential for environment management and disaster monitoring, spatial information networking in Asia and the Pacific should be given a high priority and regional cooperation in networking should be strengthened. It was recommended that NASDA and ESCAP should further promote regional cooperation in that endeavour.

Recognizing that demonstration of satellite data for public awareness was of great importance, sample satellite data illustrating successful uses of remote sensing technology should be compiled and made available to users by space agencies and interested institutions through their homepages.

Considering that the Internet had become widely accessible, efforts should be made to encourage data requests and data purchase ordering through the Internet.

Considering also that improving communication between remote sensing communities and government planners and decision makers was essential for operational use of remote sensing, it was recommended that efforts should be made to create awareness among decision makers and the public through various media such as newsletters, newspapers and television in addition to periodic high-level seminars for decision makers.

ADOPTION OF THE REPORT

The report of the Seminar was adopted on 5 November 1997 at the high-level segment. The field trip and the hands-on training continued until 7 November 1997.

The Seminar was closed by Shunji Murai, Programme Chairperson. The closing session was addressed by the Chief of the Space Technology Applications Section of ESCAP. A vote of thanks was moved on behalf of the participants by Nazmul Hoque of Bangladesh. He thanked NASDA and ESCAP for the organization of the seminar, RESTEC and AIT for technical support and the Government of Viet Nam for their local support. The organizers expressed their heartfelt gratitude to Nguyen Van Hieu and Tran Thanh Long for graciously opening the Seminar and through them to the Government of Viet Nam for hosting the Seminar.

PART TWO

PAPERS: FUTURE PROSPECTS FOR REMOTE SENSING

II. EARTH OBSERVATION IN THE MULTIMEDIA AGE

*Shunji Murai**

ABSTRACT

The Earth observation technology that started with Landsat-1, launched by the United States of America in 1972, is now entering the third generation in conjunction with multimedia technology.

Space technology applications, with not only space-borne remote sensing but also the Global Positioning System (GPS) and satellite communication, are now playing a major role in monitoring the Earth's environment, with support from geographic information systems (GIS).

Today, continuous measurement of a position with a few millimetres accuracy is possible with GPS, while real-time positioning accuracy within 20-50 centimetres has become available on board vehicles, boats and airplanes. By 1999, very-high-resolution satellite imagery with one metre ground resolution will become commercially availble. Those data will be able to be transmitted to anywhere in the world via the Internet or satellite communication.

What is more, thanks to the advancement of digital image-processing, such as animation and virtual reality, the results obtained from monitoring the Earth's environment can be represented in a more sensible form that is readily understandable.

This article reviews the present and future status of monitoring the Earth's environment in the "multimedia age".

A. Highly advanced technologies to support the multimedia age

The present age can be called the "information age", "space age" or "multimedia age", supported with electronics such as computers, optical electronics such as lasers, communication technology such as the Internet and so forth.

In the field of Earth observation or Earth sciences, a new discipline called "geoinformatics" or "geomatics" is being developed with the so-called "3S technologies": remote sensing, GIS and GPS (Murai and Fritz 1997).

Advanced technologies in geoinformatics have given rise to the following eight socio-technological changes (Fritz and Murai 1997):

- Paper maps to electronic media
- Aerial survey to space remote sensing
- Static to dynamic measurement
- Single- to multi-disciplinary professions
- Centralized operations to distributed networks
- Analogue to digital systems
- Manual to automated operation
- Closed to open data policy

* Space Technology Applications and Research Program, Asian Institute of Technology, Bangkok.

Earth observation in the multimedia age will be able to facilitate sharing of or access to data and information, thanks to the following three developments (Fritz 1997):

(a) A large volume of satellite data can be transmitted anywhere by connecting sensor platforms, sensor control stations, communication relays and portable units;

(b) Collaboration on international standards such as ISO/TC 211 or the Open GIS Consortium is being promoted to allow the sharing or exchange of geospatial data or GIS software (Tom 1997);

(c) Market forces, such as Internet access, a collaborative business environment and the trend towards privatization of mapping and charting, are now accelerated.

With these high technologies, and the standardization of a better business environment, geospatial data/information obtained from monitoring of the Earth's environment will provide users with value-added products.

B. The third generation of remote sensing technology

Remote sensing can be categorized into the following generations:

The first generation: 1972-1985. The first Earth observation satellite, Landsat-1 with multispectral scanner (MSS, 80-metre resolution), was launched by the United States of America in 1972, and Thematic Mapper (TM, 30-m resolution) was on board Landsat-4 in 1982. This generation was the American-led age when many experiments and research projects were made. Many remote sensing users demonstrated that TM data could produce thematic maps such as land-cover maps, forest maps and geological maps at 1:250,000 scale;

The second generation: 1986-1997. Following SPOT, which was launched in 1986 by the Government of France with a High Resolution Visible (HRV) scanner (10-m resolution for panchromatic with stereo function, and 20 m for multispectral), Japan launched its Marine Observation Satellite, MOS-1, with MESSR (50-m resolution) in 1987, JERS-1 with OPS (20-m resolution) and SAR (L band, 18-m resolution) in 1992, and ADEOS with AVNIR (8-m resolution for panchromatic and 16-m for multispectral) in 1996 (unfortunately, operation stopped in July 1997). India launched IRS-1A in 1988, IRS-1B in 1991 and IRS-1C in 1993 with LISS (5.8-m resolution for panchromatic and 36-m resolution for MSS). The European Space Agency (ESA) launched ERS in 1991 with SAR (C band, 30-m resolution), while Canada launched Radarsat in 1995 with SAR (resolution changeable: 9-100 m). The second generation is characterized by international participation in Earth observation for operational use. In this generation, thematic maps as well as topographic maps with contour lines at 40-m intervals, at 1:100,000 scale, were made available;

The third generation: 1998 to the future. In late 1997 and the beginning of 1998, three American private companies, namely Space Imaging EOSAT, Orbital Science and Earth Watch, planned to launch commercial high-resolution satellites with sensors that have 1-3 m resolution for PAN and 4-8 m for MSS.

The author defines the third generation of remote sensing as starting with commercialization. Better resolution, shorter revisit times and quick data distribution (within 15 minutes minimum after reception) will be available, which will be much better than the existing systems based on governmental procedures. One metre resolution satellite data will make it possible to identify individual small houses or roads, which will make possible thematic and topographic maps with 10-m contours at 1:25,000 scale.

Table 1 shows the three generations of remote sensing and their characteristics.

Table 1. Three generations of remote sensing

Generation	Purpose	Satellite/sensor/products
I. 1972-1985	Experiments and research	Landsat MSS, TM (United States) 1:250 000 thematic maps American-led technology
II. 1986-1997	Operational use	Landsat TM (United States) SPOT HRV (France) MOS-1 MESSR (Japan) JERS-1 OPS, SAR (Japan) ADEOS AVNIR (Japan) IRS LISS (India) ERS SAR (ESA) Radarsat SAR (Canada) International participation 1:100 000 thematic and topographic maps
III. 1998-future	Commercialization	Space Imaging EOSAT/CARTERRA: 1-m PAN Orbital Science/ORBVIEW: 2-m PAN Earth Watch/Early Bird: 3-m PAN Quick Bird: 1-m PAN 1:25 000 thematic and topographic maps

Source: Murai, 1997.

In the third generation, the following advanced technologies will be developed further and partly operationalized:

High-resolution satellite data. A large amount of high-resolution satellite data will be available, as shown in table 2, for providing a variety of products, such as contour maps, PAN-sharpened colour images, ortho-images, thematic maps, image overlays with GIS data, and three-dimensional landscapes (Sinclair 1997);

Multipolarization SAR. Synthetic aperture radar (SAR) is particularly useful in areas such as tropical regions where cloud cover is a serious consideration, because SAR can penetrate clouds and even see the surface at night time. SAR image quality depends on many parameters such as vertical (V) or horizontal (H) polarization for transmission and reception, frequency (L, C or X band) and off-nadir angle. In the beginning, satellite SAR such as JERS SAR or ERS SAR was available only with mono-polarization, frequency and off-nadir angle. But Radarsat adopted variable incidence angle or resolution. In the near future, multi-polarization SAR with all four possible combinations of HH, HV, VH and VV will be available;

SAR interferometry. SAR interferometry, or In SAR, is a quite new technology by which ground height can be measured by using interferometric theory, which states that two radar beams transmitted from two stations with a short base length will give the phase difference to be converted into the height between the surface and satellite. In SAR is also applicable to the detection of land deformation caused by earthquakes and volcanic eruption. Now an international project on global DEM (digital elevation model) that covers the entire world with one-second intervals in latitude and longtitude (about 30 metres on the ground) is being planned;

Imaging spectrometer. Whereas an ordinary multispectral scanner usually has a limited number of bands, ranging from 4 to 14, an imaging spectrometer has high resolution of from 64 to 384 spectral bands, which enables continuous measurement of spectral reflection from the object.

Table 2. High-resolution satellite data with less than 10-m resolution

Country and/or programme	Satellite	Data availability	Resolutions
France	SPOT	1986- available	10-m PAN 20-m MSS
India	IRS-1C	1993- available	5.8-m PAN 36-m MSS
Japan	ADEOS	1996-1997 available	8-m PAN 36-m MSS
Russian Federation	KVR-1000	Available	2-m FILM
Space Imaging EOSAT (United States)	CARTERRA	December 1997-	1-m PAN 4-m PAN
Orbital Science (United States)	ORBVIEW	1998-	2-m PAN 8-m MSS
Earth Watch (United States)	Early Bird Quick Bird	1997- 1998-	3-m PAN 1-m PAN 4-m MSS
GERS (United States)	GEROS	1998	TBD 10-m MSS
Motorola (United States)	Experimental	Not available	1-m PAN
GDE/TRACOR	Not available	1998	1-m PAN
NASA-CTA (United States)	Lewis and Clark	Late 1997	Hyperspectral 3-m PAN
IAI-CORE (Israel)	OFEQ/EROS	1997	2-m PAN
MBB (Germany)	MOMS	Not available	4.5-m PAN 13.5-m MSS

This is called a "hyperspectral imager". It will be very useful for detecting rock types, ocean colours and vegetation types. Airborne imaging spectrometers have already been developed. A spaceborne imaging spectrometer is expected to be on board *Lewis and Clark* of NASA in 1997;

Airborne laser scanner. Airborne laser scanner is used to measure the height of ground objects and man-made structures three-dimensionally, using laser beam theory, which states that the distance between the target and laser instrument can be measured by the phase difference between the transmitted and received wave. In order to determine the equation of a three-dimensional laser beam, GPS and an inertia navigation system (INS) are to be used to measure the position and attitude (roll, pitch and yaw angles) at the same time as data acquisition with the laser scanning. At present, contour maps at 5-m intervals, at 1:15,000 and 1:20,000 scale, can be provided from an altitude of 1,000 metres. The height not only of houses but also of trees can be measured;

Small satellites. The existing Earth observation satellites are expensive because the rocket size and the payload are quite large. However, compact module components such as sensors, solar panels and communication units are now available to design so-called "small satellites". Small satellites will weigh several dozen kilograms, whereas the existing large satellites are several hundred kilograms in weight. Australia, Malaysia, the Republic of Korea, Thailand and other countries and private companies are planning to launch small satellites for Earth observation, which will accelerate commercialization or privatization of remote sensing data;

Portable receiving stations. Most of the existing receiving stations are fixed at a certain place, but portable antennas and processing facilities are now available, as is the case at the Hiroshima Institute of Technology. This will facilitate better data service at a place closer to users.

C. Data processing and information extraction/production in the multimedia age

The following types of multi-function data obtained from Earth observation satellite will become available (see table 3): multispectral, multitemporal, multisensor, multiresolution, multisource, multistation, multipolarization and multimedia.

Table 3. Availability of multifunction satellite data

Type of data combination	Type of data (image) concerned	Types of processing and analysis possible
Multispectral	Optical Thermal	Hyperspectral classification
Multitemporal	Optical Thermal SAR	Change detection, trend analysis
Multisensor	Optical + Thermal + SAR	Data fusion
Multiresolution	High + low resolution images	PAN-sharpended colour composite
Multisource	Satellite images, GPS and GIS data	Spatial analysis
Multistation	Optical	Stereo vision, DEM, ortho-image
	SAR	SAR interferogram, DEM, height deformation
Multipolarization	SAR	Composite of HH, HV, VH and/or VV, classification
Multimedia	All types	Communication, networks, visualization

The total amount of NOAA AVHRR data received in one year adds up to about 1 terabyte, whereas the total amount of satellite data received by NASA in one day reaches 12 terabytes. Such a huge amount of data should be not only stored and processed but also used to extract useful information.

In the multimedia age, with the ability to communicate data/information to users in any place in a very short time, the following software developments for data processing and information extraction/production should be targeted:

(a) Improvement of the accuracy of classification with multi-function satellite data such as hyperspectral analysis, multitemporal analysis, data fusion of multi-sensor data and other techniques;

(b) Development of special techniques to produce cloud-free data, such as the use of NOAA AVHRR every 5 or 10 days;

(c) Development of a so-called "image understanding technique" to recognize man-made structures automatically and to input them to a GIS database, which would be a great contribution to improving GIS database management;

(d) Automatic interpretation of not only man-made structures but also natural features, which would make it possible to produce maps automatically by combining the stereo matching technique to generate digital elevation models;

(e) Data fusion of multisensor and multiresolution satellite images since more information can be extracted with data fusion techniques to merge optical and SAR data, or high-resolution panchromatic and low-resolution multispectral data;

(f) The use of three-dimensional GIS, i.e. stereo imagery, laser scanner and SAR interferometry, to generate three-dimensional surface models; in particular, for urban 3-D structures, new developments in 3-D urban GIS are needed;

(g) Development of Earth environment monitoring and prediction on a global scale using global data sets obtained from Earth observation satellites. In particular, the assessment of the impact on humans of population increase, deforestation and land degradation, is now an urgent issue in global change analysis and prediction;

(h) In addition to virtual reality or animation techniques, visualization of geospatial data or information taking into consideration human sensibility is now important for computer mapping. Physical, physiological and psychological aspects will be considered in creating new virtual reality maps.

D. Conclusions

On the basis of the above, the author would like to share his vision of the future, which is as follows:

(a) Monitoring the Earth's environment by remote sensing will be more and more commercialized, with many competitive functions, such as high resolution, hyperspectral analysis, frequent revisits and quicker data distribution;

(b) Software with artificial intelligence and multimedia techniques will be further developed, resulting in improved information extraction and production in actual applications. More high-value-added information will be produced by integrating remote sensing and GIS with visualization based on human sensibility;

(c) Information about the latest status of and predicted changes in the Earth's environment will become available anytime and anywhere in the multimedia age.

Bibliography

Fritz, L. and S. Murai, 1997. Activities of the International Society for Photogrammetry and Remote Sensing (ISPRS). The United Nations Regional Cartographic Conference on Asia and the Pacific, Bangkok (February).

Fritz, L., 1997. Technology developments for the Spectral Information Community. Paper delivered at the 15th International Hydrographic Organization (IHO) Congress.

Guo, H., 1997. Spaceborne multi-frequency, polarametric and interferometric radar for detection of targets on Earth surface and subsurface (in Chinese), *Journal of Remote Sensing,* 1(1):32-39.

Murai, S. and L. Fritz, 1997. Integration of 3S technologies: remote sensing, GIS and GPS. The United Nations Regional Cartographic Conference on Asia and the Pacific, Bangkok (February).

Murai, S., 1997. Advanced technologies and their prospects in geoinformatics (in Japanese), *Nikkei Construction* (23 May), pp. 75-81.

Sinclair, S., 1997. Satellite for GIS infrastructure planning and management. In *Proceedings* of IEAS'97 and IWGIS'97, Beijing (August), pp. 601-610.

Tanaka, N., K. Ono and S. Murai, 1997. A new map representation in consideration of sensibility (in Japanese). Workshop on New Technologies in Geomatics, Tokyo (September).

Tom, H., 1997. The global spatial data infrastructure (GSDI). In *Proceedings* of IEAS'97 and IWGIS'97, Beijing (August), pp. 28-38.

Tong Qinxi and others, 1997. Study on imaging spectrometer remote sensing information for wetland vegetation (in Chinese), *Journal of Remote Sensing,* 1(1):50-57.

III. FUTURE PROSPECTS FOR MAPPING FROM SPACE

*Gottfried Konecny**

Mapping from space has become possible through the phenomenal development of space platforms and space sensors during the past generation.

Mapping from space may be considered a technology-driven activity, but it is vitally needed for the provision of basic information required for sustainable development.

A. Global economic development and technical cooperation

Human activity is based upon economic development. Throughout human history economic development has gone through four different stages:

Nomadic -----> Agricultural -----> Industrial -----> Service oriented

Owing to different conditions in different parts of the world influenced by facters such as climate, soil conditions, mineral resources, labour, education, technical innovation and motivation, this economic development has progressed at different rates in different parts of the world.

The United Nations statistical yearbooks list a number of parameters, according to which this progress is usually measured in the countries of the world:

Percentage of employees in agriculture, industry and services;
The gross national product (GNP) or gross domestic product (GDP) per inhabitant;
Percentage of food supply;
Inhabitants per medical doctor;
Child mortality.

The countries of the world can usually be divided into three categories:

(a) Low-income countries characterized by annual GNP under US$ 600 per inhabitant, a predominance of agricultural economy, and a shadow economy of over 50 per cent. These countries have recently shown a decline in the GNP;

(b) Medium-income countries with a GNP between US$ 600 and 3,000 per inhabitant, a predominance of industrial activity and a shadow economy under 20 per cent. Among these may be included the socialist reform countries with a stagnating GNP, the "tiger" countries of Asia with high foreign investment and the highest GNP growth rates, and the debtor countries, mainly of Latin America, with stagnating GNP growth rates;

(c) The high-income countries with a GNP of over US$ 3,000 per inhabitant. They are service oriented, and their shadow economy is less than 10 per cent. Among these may be counted the donor countries with small GNP growth rates and the oil-exporting countries with stagnating GNP growth rates.

One of the major achievements of the United Nations system has been to stimulate high-income countries to share some of their wealth with the countries of lower income to stimulate their economic development through technical cooperation.

The following four countries alone account for over 50 per cent of the total economic cooperation of US$ 58 billion/year, for example, available in 1991:

* Institute for Photogrammetry and Engineering Surveys, University of Hannover, Germany.

	US dollars
United States of America	11.3 billion
Japan	11.0 billion
France	9.5 billion
Germany	6.8 billion

Most of the development funding went to sub-Saharan Africa (32.5 per cent), but a significant portion also went to South-East Asia (27.1 per cent).

There have been many mutual discussions to make sure that some of the difficulties encountered, particularly in the poorest part of the world, Africa, can be overcome.

The difficulties stem from institutional, financial and educational difficulties in the countries, but they sometimes also relate to unsuitable technical issues.

B. Global trends and sustainable development

The major trend in economic development is due to population growth. The current population of the world is 6 billion people, and this figure will most likely double in the next 50 years.

Because of this trend, there is additional need for food production and a need for sustainable development to preserve the global ecosystem with respect to a sustainable water balance, the mitigation of drought and the preservation of coastal waters.

This leads to the necessity to monitor degraded forests, poor crop yields, dumps, drought areas, floods, sedimentation, soil erosion and desertification, and the growth of urban areas.

The United Nations Conference on Environment and Development, held at Rio de Janeiro, Brazil, in 1992, in Agenda 21, chapter 40, clearly describes the monitoring needs.

Another conference, the United Nations Conference on Human Settlements (Habitat II), held in Istanbul, Turkey, in 1996, clearly pointed out that rapid population growth occurs mainly in the poorly developed countries, with 2.7 per cent growth per year, and that it goes hand in hand with urbanization of up to 5 per cent per year in these countries (see table 1).

The urban population is at present about 45 per cent of the total; 80 per cent of the world population is expected to be urban by the year 2025.

Most of the highest urban growth rates are expected to be in Asian cities, such as Dhaka (7.8 million) 5.7 per cent, Jakarta (11.5 million) 4.4 per cent, Karachi (9.9 million) 4.3 per cent,

Table 1. Population statistics and predictions

Continent	Population 1996 (millions)	Estimated population 2025 (millions)	Growth 1995-2000 (percentage)	Urban population (percentage)	Urban growth 1995-2000 (percentage)
Africa	784	1 496	2.7	34	4.3
Asia	3 513	4 960	1.5	35	3.2
Europe	727	718	0.1	74	0.5
Central and South America	490	710	1.7	74	2.3
North America	296	370	0.9	76	1.2
Oceania	29	41	1.4	70	1.4
World	5 804	8 294	1.5	45	2.5
Developed countries	1 171	1 238	0.3	75	0.7
Countries under industrialization	4 633	7 056	1.8	38	3.3
Poorly developed countries	592	1 162	2.7	22	5.2

Mumbai (15.0 million) 4.2 per cent, Shanghai (15.1 million) 2.3 per cent and Bangkok (7.1 million) 2.2 per cent.

C. The need for basic data sets in geographic information systems

In the last 20 years geographic information systems (GIS) have been developed as computer systems capable of input, storage, manipulation, analysis and output of geographic data.

In its wider definition a GIS is, however, a system used in managing the environment for sustainable development by providing data for the following purposes:

Gaining information through the analyses of data;
Planning;
Decision-making;
Implementation and monitoring of decisions.

The hardware and software of a GIS rarely exceeds 20 per cent of its cost, while the data account for 80 per cent of the cost. Moreover, the data system needs to be kept up to date.

A GIS, thanks to its data integration capability from various sources, has the advantage of being at least four times more cost-effective than the simple computer automation of a task.

D. The need to provide timely base data sets at various scales

A survey of data acquisition costs for various purposes shows a scale dependence. The larger the scale, the more costly the data. Costs of less than US$ 100 per square kilometre can be achieved only with satellite imagery. Aerial or ground survey tools, which can supplement such surveys, are in general at least 10 times as costly per square kilometre (see table 2).

Mapping and GIS consist of a base map coverage with integrable thematic layers and references to non-graphic data in databases in table form (see figure I).

It is this interpretation of information that makes efficient data management possible and affordable.

E. The need for satellite data systems to provide the data

The United Nations Secetariat has tried to monitor existing base map data for the different countries and continents at different scales (see figure II). The result has been that global coverage exists only at scales smaller than 1:250,000. At 1:50,000 about two thirds of the land area is covered, and at 1:25,000 about one third. What is even more alarming is that the present update rate for the 1:50,000 map is only 2.3 per cent and that of a 1:25,000 map 4.9 per cent. So the average age of a 1:50,000 map is 45 years and that of a 1:25,000 map 25 years (see figure III). The continents of South America, Australia and Oceania, Asia and Africa have much smaller (i.e. longer) update rates than Europe or North America. It becomes clear that the existing map technology based on aerial photography and ground methods is too slow to provide the required data sets. Therefore, satellite systems must be utilized.

F. Present capabilities of optical satellite systems

Geostationary low-resolution satellite such as GMS, Insat, GOES and Meteosat offer images of the Earth's surface every 30 minutes at 5-km ground pixels. NOAA satellites offer 1-km resolution at least twice per day. Such data are ideal for global monitoring.

Resource satellites such as Landsat, SPOT, JERS and IRS-1A and 1B offer medium-resolution data between 10- and 30-m ground pixels several times per year.

Table 2. Survey costs

Field	Type	Scale	Imagery	Cost/sq km (US dollars)
Agriculture	Phenol change	1:1 000 000	NOAA	80
Bio-material	Biomass change	1:1 000 000	NOAA	80
Forestry	Forest mapping	1:250 000	MSS	6
Geology	Reconnaissance	1:100 000	TM	20
Forestry	Forest development	1:100 000	TM	20
Irrigation	Watershed mapping	1:100 000	TM	10
Regional planning	Planning study	1:100 000	TM	25
Land use	Land use mapping	1:100 000	TM	13
Bio-material	Biomass inventory	1:100 000	TM	20
Erosion	Vegetation cove	1:100 000	TM	20
Desertification	Change detection	1:100 000	TM	35
Food security	Cultivation inventory	1:100 000	TM	25
Environment	Environment inventory	1:100 000	TM	50
Regional planning	Feasibility study	1:50 000	SPOT XS	40
Environment	Risk zone mapping	1:50 000	KFA 1000	150
Urban development	Urban change	1:50 000	KFA 1000, SPOT PAN	45
Topography	Base map	1:50 000	Aerial photo	120
Geology	Photogeology	1:25 000	Aerial photo	150
Transport	Road design	1:20 000	Aerial photo	180
Topography	Orthophoto	1:12 000	Aerial photo	24
Water supply	Base map	1:10 000	Aerial photo	800
Forestry	Forest inventory	1:10 000	Aerial photo	350
Land use	Land use mapping	1:10 000	Aerial photo	520
Bio-material	Energy study	1:10 000	Aerial photo	250
Transport	Photographic map	1:10 000	Aerial photo	700
Cadastre	Orthophoto map	1:10 000	Aerial photo	400
Topography	Base map	1:5 000	Aerial photo	2 000
Topography	Orthophoto	1:5 000	Aerial photo	78
Cadastre	Photographic or survey map	1:2 000	Aerial photo	10 000
Cadastre	Orthophoto	1:2 000	Aerial photo	1 000
Topography	Orthophoto	1:1 000	Aerial photo	800
Urban cadastre	Base map	1:1 000	Aerial photo	20 000
Urban cadastre	Multipurpose cadastre, utilities, topography	1:500	Aerial photo	40 000

The latest development is high-resolution satellites such as IRS-1C and MOMS, with about 5-m ground pixels, but without the capability to obtain global coverage as yet (see figure IV).

Experience with these systems has shown that the use of satellite images for mapping is at least four times cheaper than using conventional methods, but that at present resolutions quality standards must be relaxed; visual interpretation of these images is still more effective; but GIS integration is of advantage; cloud cover is still a handicap, but possibilities are open for radar satellites such as JERS-1, ERS-1 and 2 and Radarsat.

Even 5-m resolution systems cannot compete in quality with aerial photography with 1-m resolution in 1:25,000-scale mapping (see figure V).

The suitability for mapping from a particular type of imagery has already been investigated, in the early 1980s. There are three criteria to be met:

(a) Planimetric accuracy, which is scale dependent;

(b) Elevation accuracy, which depends on parallaxes created by the different image geometry from two different imaging positions;

(c) Detectability, which relates to the spatial resolution, which may be achieved by a particular sensor system.

Even though aerial photography, which has been digitized into different pixel sizes on the ground, can hardly distinguish more than six bits of grey values, as opposed to recent digital sensors with 10 or more bits of grey level distinction, these early results are still generally valid.

Planimetric accuracy of a map is generally related to ± 0.2 mm at publishing scale according to United States mapping standards. This criterion mainly relates to worldwide mapping practices for the original mapping scale, but not for generalized maps at smaller, derived scales, in which the planimetry is often shifted to accommodate conflicts in the depiction of objects. Even in the original mapping scales, buildings or building blocks, roads and rivers are shifted in some national map bases for this purpose. But, in general, this means from the data acquisition side, that the following planimetric standards are usually accepted:

Scale	σp
1:10 000	± 2 m
1:25 000	± 5 m
1:50 000	± 10 m
1:100 000	± 20 m
1:200 000	± 40 m

Elevation accuracy is generally a function of terrain slope. Depending on terrain slope, a certain contour interval is specified. The reliability of contouring is generally accepted as being five times the point measuring accuracy in height, regardless of whether the contours are originally measured in a photogrammetric plotting instrument, or whether they are interpolated on the basis of a measured digital elevation model (DEM) grid.

Contour interval Δh	Point measurement accuracy $\pm \sigma h$	Terrain type
1 m	± 0.2 m	Flood plain
2 m	± 0.4 m	Flood plain
5 m	± 1 m	Flood plain
10 m	± 2 m	Flood plain
20 m	± 4 m	Flood plain
50 m	± 10 m	Flood plain
100 m	± 20 m	High mountains

The detectability of objects, given sufficient contrast as a function of grey level discrimination, was formerly measured in terms of photographic resolution stated as line pairs per millimetre (lp/mm). Nowadays this photographic resolution must be compared to 2 to 5 pixels at image scale related to the instantaneous field of view (IFOV) on the ground.

Early tests with photographic resolutions have been carried out for specific objects to be recognized and identified from the imagery. They established a minimum pixel size for the detectability of the following objects:

Object	Pixel size
Urban buildings	2 m
Footpaths	2 m
Minor road network	5 m
Fine hydrology	5 m
Major road network	10 m
Building blocks	10 m

The highest resolution from space that is currently accessible on the civilian market is Russian KVR 1000 photography, digitized to 2-m pixels, in which individual houses can be clearly identified (see figure VI).

Such images may be used for the creation of planimetric image maps at the scale of 1:10,000, but the lack of a large number of ground control points to differentially rectify the image maps (on account of image deformation to planimetric map accuracy standards at that scale) renders the expected planimetric accuracy closer to the 1:25,000 level.

Moreover, the height determination from images generally depends on the height-base ratio of the imagery flown:

$$\sigma_h = \pm \frac{h}{b} \cdot \frac{h}{c} \cdot \sigma_{px} \qquad (1)$$

where

σh	=	Point error in elevation
h	=	Orbital height
b	=	Orbital base
c	=	Principal distance (focal length) of the camera objective
σpx	=	Parallax measurement error in the order of the point positioning error σp of about 10 μm in a photographic image, corresponding more or less to the pixel size on the ground with the image scale used.

Owing to the very long focal length of the KVR camera (1 m), the stereoscopic overlap conditions of that imagery will not permit a smaller and more favourable height-base ratio than 10, rendering the expected height accuracy less than ± 20 m.

The present KVR 1000 high-resolution images are therefore suitable for planimetric map updates of 1:10,000-scale or 1:25,000-scale maps, but not for digital elevation model measurements.

Cartographic satellites that permit a better height determination, even if they do not reach the same detectability, are the French panchromatic SPOT system, with 10-m pixels, and the Indian IRS-1C system with 5.8-m pixels. Thanks to their ability to incline the sensor by a mirror in cross-track direction, a favourable height-base ratio of up to 1 may be achieved from subsequent orbits. This ability, however, is often handicapped because of changing cloud cover, which severely limits stereoscopic coverage for a time period in which the radiometry of the ground has not changed.

The best test result achieved thus far with SPOT PAN and IRS-1C stereo imagery is in the order of ± 5 to 10 m in elevation and ± 3 to 5 m in position, making these sensors suitable for 1:50,000 to 1:100,000-scale mapping in mountainous areas.

Another approach has been provided by the in-flight stereo capability of the German Stereo-MOMS system. It consists of a triple line scanner looking forward, vertically down and aft, which was flown for 10 days on the United States-German Space Shuttle mission D2 in 1993, and which since 1996 has operated on the Mir space station's Priroda module (with interruptions).

On Mir, the vertical sensor yields 5-m panchromatic ground pixels and/or 15-m multispectral ground pixels, and the fore and aft sensors give 15-m panchromatic ground pixels. Figure VII shows such an image over the German city of Augsburg, in which the vertical panchromatic 5-m pixels have been fused with the vertical multispectral 15-m pixels.

Figure VIII shows an oblique view of three processed fore, down and aft panchromatic stereo images after the generation of a DEM and the resulting ortho-image over the area of Barcelona, Spain.

The advantage of in-track stereo sensing is that all images used to create digital terrain model - orthophotos and oblique views - are taken at the same time.

The cartographic requirements for these images show that planimetric requirements can be met for the 1:25,000 scale, that elevations can be determined with ± 5 to 10-m accuracy and that the detectability is suitable for 1:50,000 mapping.

G. Present capabilities of microwave satellite systems

Optical satellite systems have so far shown difficulties with marginal resolution and with cloud coverage. For this reason the European Space Agency (ESA) has launched the European Remote Sensing (ERS) radar satellite series, followed by Japan with JERS and Canada with Radarsat.

Radar signals follow a very different geometry from optical devices. The existing satellite systems have been designed not for land but for ocean coverage, giving differences in the amount of foreshortening or shadows to be encountered. Moreover, radar backscattering does not behave like optical reflection. So the criteria for planimetric accuracy and for detectability of objects in radar images do not easily correspond to pixel size resolution of optical systems. Depending on the grey level differentiation desired, radar processing can be executed (for example, for ERS-1) to 12.5-m or 25-m pixels related to the ground.

Radar images are therefore more suitable for image fusion with optical images rather than for mapping alone.

Radar, when used in multitemporal mode or multipolarization mode, on the other hand, compares with optical multispectral differentiation capabilities, when classifying area objects. Yet satellite radar has a very distinct advantage: it generates coherent radiation. Coherent radar signals influence radar geometry since the images are reconstructed using phase and Doppler information. They can properly be reconstructed only if a DEM is used in the reconstruction.

On the other hand, phase information may be utilized for DEM generation using interferometry principles. While radar interferometry has been carried out using repeated images from quite different orbits (as for JERS-1 and Radarsat), as long as there is a base between the two imaging stations, the conditions for interferometry were greatly improved using the ERS-1/ERS-2 tandem mission, which directed the two ERS satellites so that corresponding radar images over the same area were obtained one day apart, with a small base between orbits.

Radar interferometry requires not only that the base between orbits is known, but that the sensor position and the sensor attitude is accurately determined. In the present satellites systems such possibilities have not been available with sufficient accuracy.

In the Hannover area, a digital elevation model generated from radar interferometry has been fitted by a seven-parameter transformation to the few identifiable common controls in the radar images. In most areas the comparison with an existing and accurate conventional DEM resulted in average discrepancies of less than 10 m for 90 per cent of the points, but in hilly areas discrepancies of over 100 m were encountered.

This confirms that radar interferometry may be useful in differential changes of image portions, which are due to dynamic changes of the terrain, but it so far fails in reliable elevation determinations.

An improvement of this situation has been suggested by the United States - German SIR-C Space Shuttle mission in 1999, which not only will use three radar frequencies (X band, C band and L band) to resolve atmospheric ambiguities, but also will have two types of receiving antennas mounted on a 30-m long beam, the positions and attitudes of which are continuously monitored by differential GPS. In this way it is hoped to obtain a system capable of rapid DEM mapping for large areas to ± 10-m accuracy.

H. Digital elevation models

There are currently at least six alternative methods considered for future generation of digital elevation models, competing in accuracy and price:

(a) The digitization of existing line maps at 5- to 20-m accuracy, depending on the original map quality, at a price of US$ 1/sq km;

(b) The existing United States military global coverage at 20-m accuracy, obtainable for some portions of the globe, at a price of US$ 1/sq km;

(c) Aerial photogrammetric mapping at 5-m accuracy for US$ 40/sq km;

(d) The future use of American commercial optical satellites, following the Stereo-MOMS principles, but at higher resolution, with 5-m accuracy for US$ 50/sq km with the advantage to map areas, when aerial photography is not available;

(e) Optical stereo sensors following the Stereo-MOMS or the SPOT/IRS-1C principle with 10-m accuracy at US$ 5/sq km;

(f) Interferometric SAR of the SIR-C type with 10-m accuracy at US$ 5/sq km.

I. Use of satellite remote sensing systems

The uses of existing satellite images for monitoring the environment in the largest sense are manifold. The choice of imagery is always a compromise between availability and spatial, spectral and temporal resolution, considering repeatability, swath and pixel size.

All remote sensing conferences in Asia and in other parts of the world show that weather and meteorological dangers (storms and cyclones for instance) can be monitored by global satellites such as Insat. NOAA satellites permit scientists to measure sea surface temperature, pigment and chlorophyll concentration of ocean and coastal areas. But they are also able to monitor the state of vegetation on the basis of the normalized difference vegetation index to follow patterns of drought or floods, the health of tropical forests or their devastation by forest fires.

Resource satellites are useful in monitoring crop patterns to detect vegetation diseases, to determine erosion risks or to follow uncontrolled growth of industrial activity and urban settlements and their pollution effects on the environment.

There would be no hope for gathering the necessary amount of this type of information without satellite remote sensing systems.

These interpreted results, together with socio-economic data, constitute the needed thematic information which requires reference to the base mapping system in the form of a GIS.

Noteworthy are the recent activities of the Committee on Earth Observation Satellites (CEOS), an organization of space agencies formed on the initiative of the G-7 group. CEOS has promoted the creation of an information locator system (ILS), which helps the user to find pertinent information via the Internet. This CEOS-ILS will contain types, locations and times of satellite sensor images taken, if possible with reduced content "quicklooks". In addition, value-added products such as digitized maps and metadata of various kinds are to be located in the system to offer the user a more complete information potential.

J. Future capabilities of medium- and high-resolution sensors

There will be many more satellite systems available in the near future from many nations. ESA divides them into scientific "explorer missions" and operational "Earth watch mission". ESA, for example, plans at least four explorer missions in the 2003 to 2011 time-frame; the missions will be directed towards the following scientific goals:

(a) To measure the Earth's gravity field from space;

(b) To obtain quantified values for Earth radiation;

(c) To explore the spectral capabilities of image spectrometry in a land mission;

(d) To determine missing parameters of the Earth's atmosphere through cloud profiling and the measurement of the wind field.

These missions will enable scientists to develop better models for climate, atmosphere and other physical parameters, which could help to fill in gaps in scientific understanding. The Earth watch missions plan to improve the current capabilities of resource and cartographic satellites. Among these are the Japanese, Indian, European, and American missions ALOS, IRS-1D, Envisat, SPOT-5, and Landsat-7. But of primary interest are the American commercial ventures for high-resolution imagery.

As is usual with planned systems, the details about these ventures change almost every month. Table 3 provides a summary, compiled by L. Fritz in October 1997, of Earth observation satellites and relevant parameters.

These missions will vastly improve the present sensing capabilities, using a novel approach which will comprise an end-to-end data provision system, including corrections for calibration and reference systems, cataloguing, value-added processing and distribution.

Satellite imaging and processing capabilities may become serious competitors of the traditional aerial survey industry unless the two approaches are merged and used in supplementation. Conceptually and on the experimental level, mapping from space is now a reality, soon to become operational on a competitive basis.

K. Conclusion

On the basis of the global scenario of development from the historical perspective, the following conclusions can be drawn:

(a) The nineteenth century was the century of interference and control between nations, a time when colonial powers used to introduce their limited mapping systems for their own resource exploitation by inadequate means;

(b) The twentieth century became the century of independence and competition between nations. In this century mapping for national resource management became possible through the world-war-proven aerial photogrammetric techniques propagated through the United Nations, often with the help of donor countries;

(c) The twenty-first century is likely to become the century of interdependence and cooperation between nations. In satellite remote sensing there is hope in the coordination activities of CEOS in planning satellites for global and regional needs. There is also hope in the recent formation of international consortia to build sensing systems for satellites and to process these products as GIS input in an end-to-end system. In this scenario mapping from satellites constitutes a contribution to the preservation of living conditions on this planet by providing the necessary information for it.

Table 3. Commercial Earth observation satellites

Systems	Earth Watch “Quick Bird”	Orbital Sciences “Orb View 3”	Space Imagery “Ikonos”	West Indian Space Ltd. “EROS”	Earth Watch “Early Bird”	Resource 21 “Resource 21”	GEROS	Kodak “Cibsat”
Partners	Ball Hitachi Telespazio MDA	Orbital Sciences	Lockheed Martin, E-Systems Mitsubishi	Israeli Aircraft Industry, Core Software Technology	Ball Hitachi Telespazio MDA	Boeing Farmland GDE ITD	Geophysical and Environmental Resource Corporation, Space Vest	
Launch	1. 1998 2. 1999	1998	1. December 1997 2. September 1998	1. 1998 to 6. 2003	1997	1. 2000 (2) 2. 2001 (2)	1. 1999 (2) 2. 2000 (2) 3. 2001 (2)	1999
Mode	PAN MS	PAN MS	PAN MS	PAN	PAN MS	MS	PAN MS	PAN MS Hyper-spectral
Quantization	11 bit 11 x 4 bit	8 bit 8 bit	11 bit 11 bit	10 bit	8 bit 8 x 3 bit	12 bit		11 bit 11 bit
Resolution	0.82 m 3.28 m	1 & 2 m 4 m	0.82 m 4 m	1.3 m	3.2 m 15 m	10 m 20 m 100 m	10 m	
Channels	1 4	1 4	1 4	1	1 3	4 2 1		1 5 60
Swath	22 km	8 km	11 km	13.5 km	6 km 30 km	205 km		112 km
Pointing in track	± 30°	± 50°	± 45°	± 45°	± 30°	± 30°	–	2 convergent sensors
Pointing cross track	± 30°	± 50°	± 45°	± 45°	± 28°	± 40°	–	–
Sensor position	GPS	GPS	GPS	GPS	GPS	GPS	GPS	GPS
Sensor attitude	Star trackers	2 star trackers	3 star trackers	–	1 star tracker	Star trackers	Star trackers	2 star trackers
Expected accuracy with GCPs	Horiz. 2 m Vert. 2 m	Horiz. 7.5 m Vert. 3.3 m	Horiz. 2 m Vert. 3 m	Horiz. 6 m Vert. 4 m	Horiz. 6 m Vert. 4 m	5 m abs. 1 m rel.	Horiz. 3 m	Horiz. 5 m Vert. 3 m
Without GCPs	23 m 17 m	12 m 8 m	12 m 8 m	800 m –	150 m –	30 m	25 m	– –

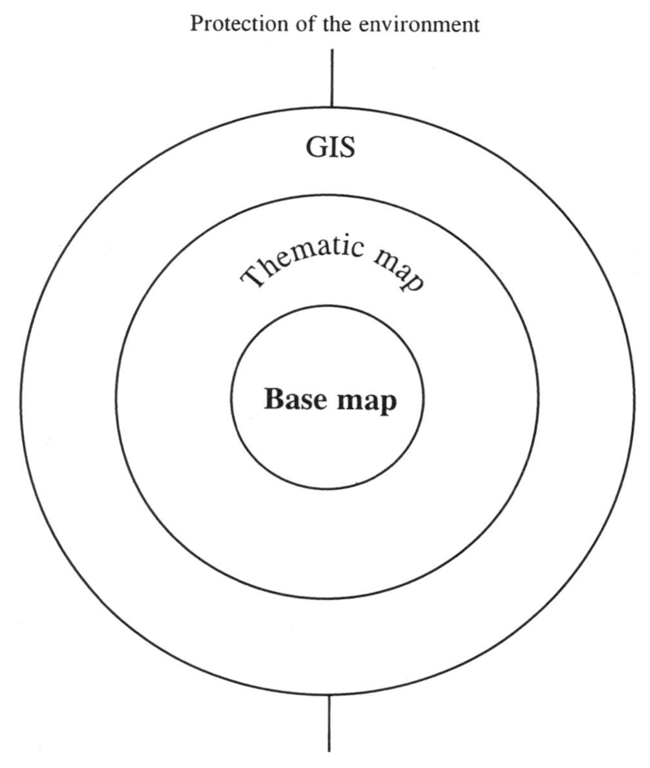

Figure I. Mapping and GIS

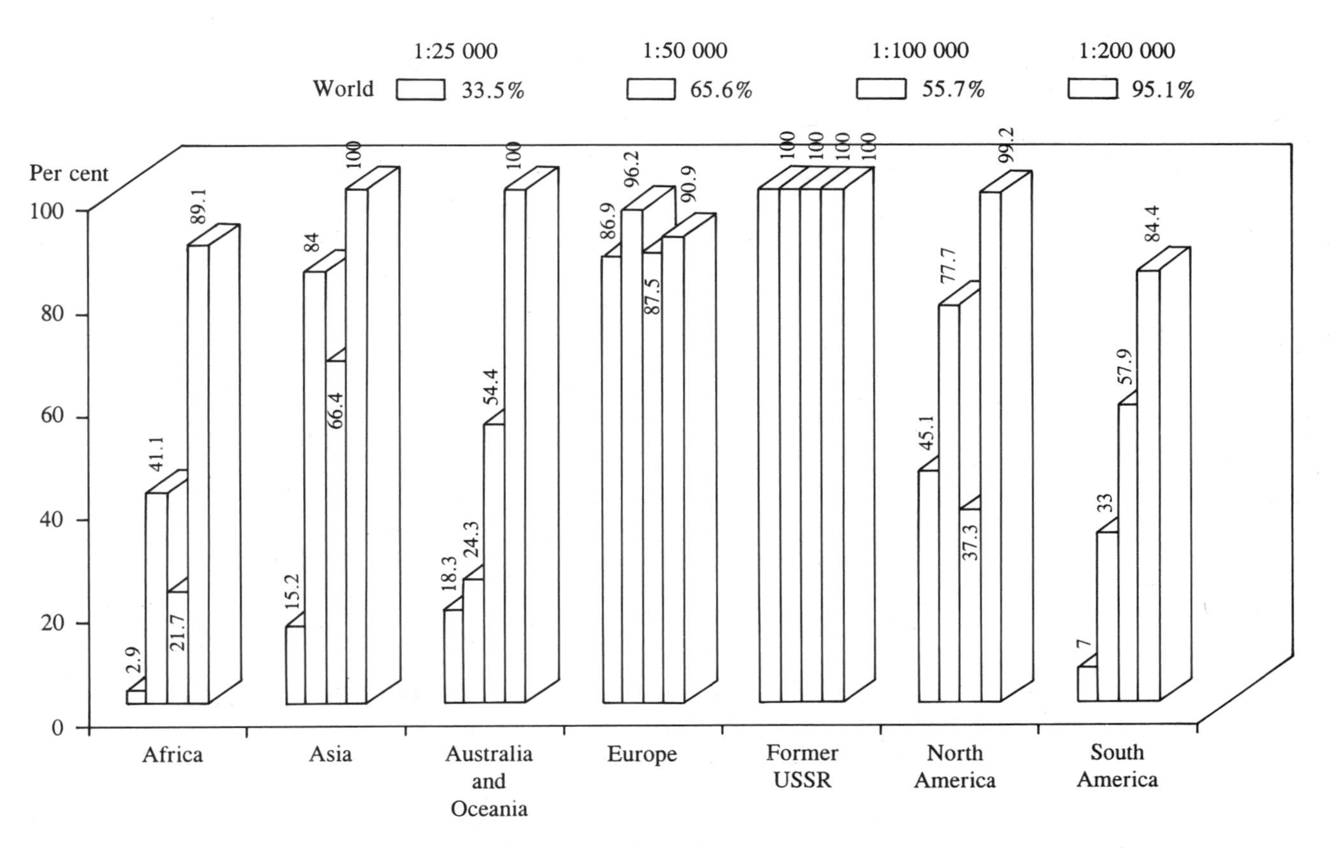

Source: United Nations Cartographic Conference, Beijing, 1993.

Figure II. Status of mapping (United Nations Statistics of 1990)

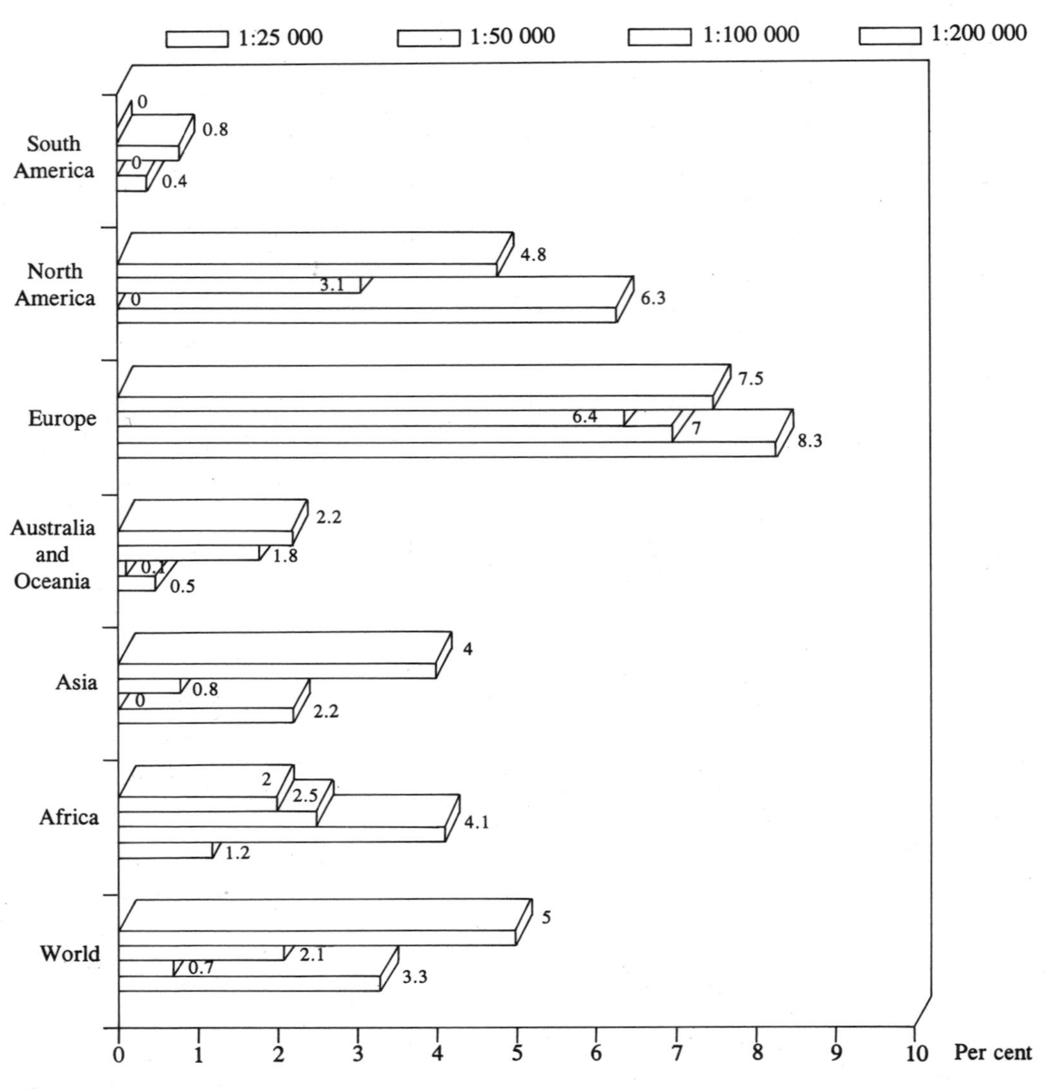

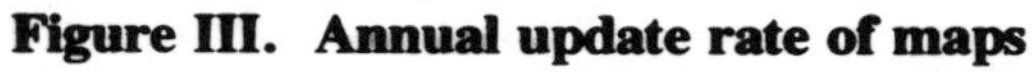

Figure III. Annual update rate of maps

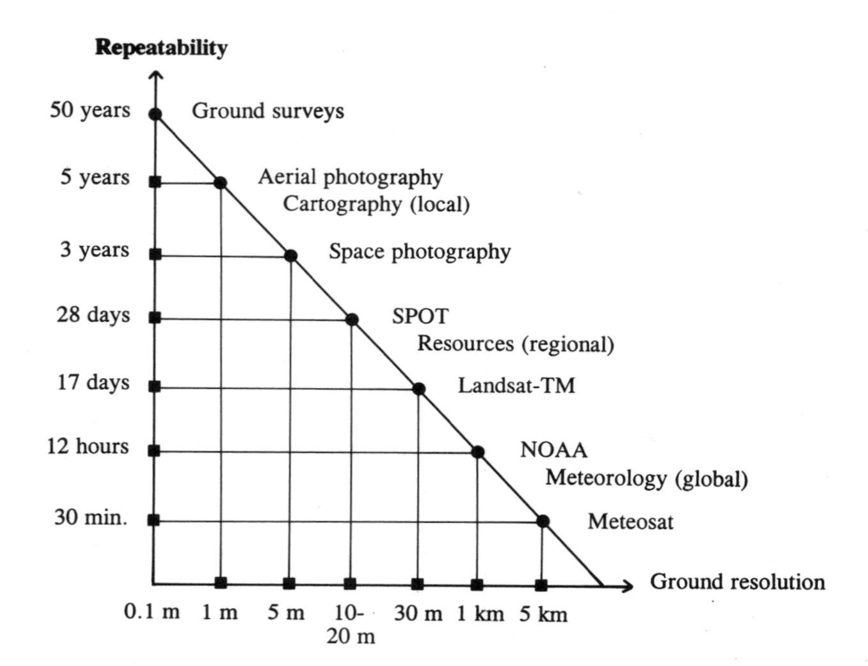

Figure IV. Resolution and repeatability of remote sensing systems

a. SPOT multispectral

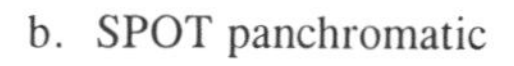

b. SPOT panchromatic

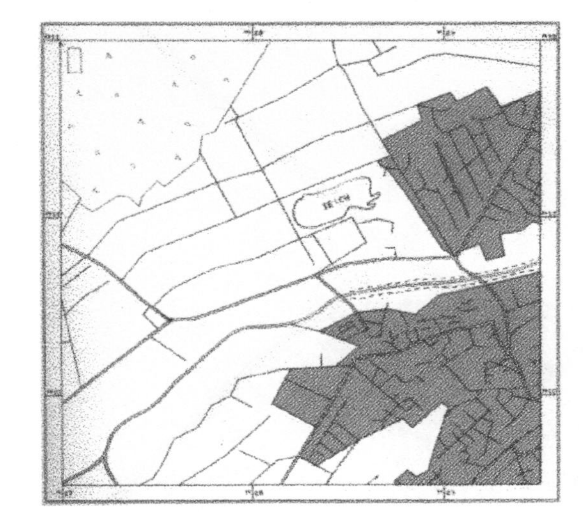

c. KFA 1000

d. High-altitude photo 1:20,000

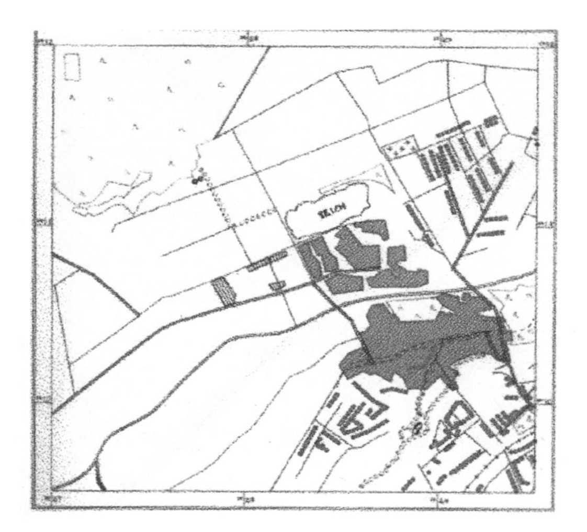

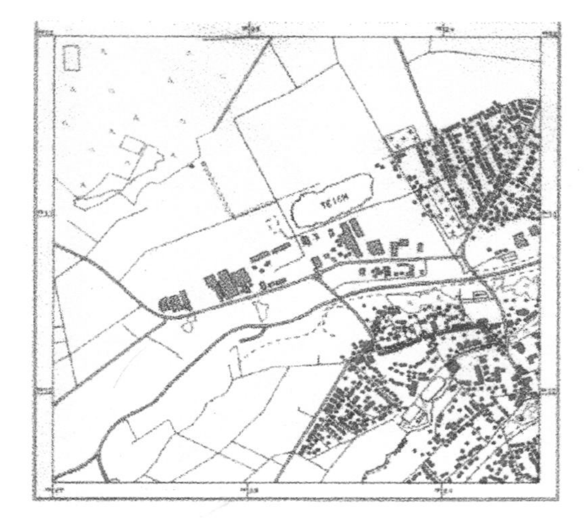

e. TK 25

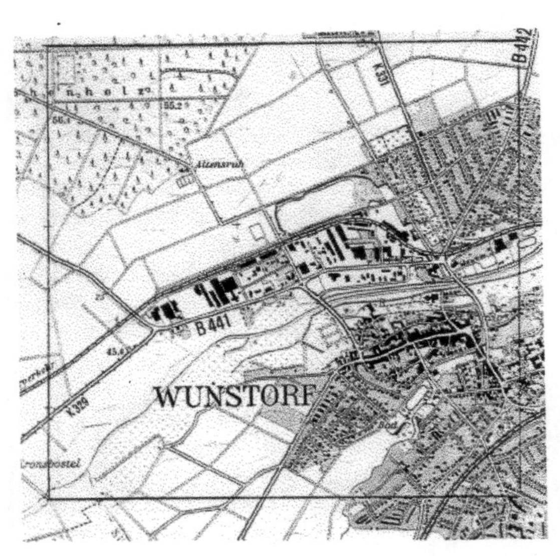

The boundaries and names shown and the designations used on this map do not imply official endorsement or acceptance by the United Nations.

Figure V. Mapping from satellite imagery from a. SPOT MS (France), b. SPOT PAN (France), c. KFA 1000 (Russia), d. aerial photography 1:20,000, and e. existing map 1:25,000

Figure VI.
Digitized KVR 1000 image (DD5) with 2-m pixels over the city of Hannover, Germany

Figure VII.
MOMS-Priroda image over Augsburg, Germany (5-m pixels)

Figure VIII. Oblique view of differentially rectified MOMS-Priroda stereo images over Barcelona, Spain

IV. JAPANESE EARTH OBSERVATION PROJECT

*Takashi Moriyama**

A. Importance of tropical ecosystems in Earth environment change

Tropical ecosystems are vulnerable to:

- Earth environment change (long term)
- Climate variability (seasonal, annual)
- Natural disasters (unexpected)

As the number of tropical rainforests decreases, estimated CO_2 sink capability might be changed and thus influence the balance of global CO_2 circulation in the Earth system. This change makes it difficult to estimate the trend of global warming. Biomass burning by wildfires should be reduced. NASDA has been developing a global rainforest map using JERS-1 data. NASDA will make efforts to update this map and provide it to users through joint research, pilot projects and so on, free of charge.

B. Satellite measurement contribution: Japan's plan for the near future

The most frequent satellite measurements are used for long-term monitoring and global coverage. In particular, high spatial resolution images in visible, near-IR and SWIR are the most appropriate data, generally for land surface applications. Currently, Landsat, SPOT, MOS, JERS-1, ADEOS and IRS have been used, and advanced sensors aboard Landsat-7, SPOT-4, ADEOS-2 and ALOS are planned for launch.

Since tropical rainforest regions are usually covered by thick cloud, SAR is convenient to use combined with optical data by adopting data fusion technology. For the users, satellite data matched up with ground truth are essential. For the purpose of complementary use of the satellite data, this means the data do not depend on a single satellite; data fusion technology is very important to promote multiple sensor data utilization.

C. User involvement, support and outreach

As for the international approach, the Committee for Earth Observation Satellites (CEOS) is the major body that can accomplish this through its activities. CEOS has three working groups: (a) AG/IGOS, to make the connection between user requirements and sensor development by space agencies, and to promote international cooperation by exchanging/sharing missions, sensors and satellite development; (b) CAL/VAL, to ensure satellite data quality and standardized CAL/VAL methodology; and (c) WGISS to promote user involvement and support, especially for developing countries, by facilitating the data access and adopting the developed standardized method, such as CILS (CEOS Information Location System), CII (Catalogue Interpretability Infrastructure) and others. WGISS developed a five-year plan, in which the involvement of developing countries (especially in South-East Asia) has the highest priority.

Harmonized with CEOS activities, NASDA is keen to promote the following actions:

- Data utilization training
- International workshops, seminars

*National Space Development Agency of Japan.

- Pilot projects
- RA (research announcement) for Asia
- CILS installation
- Data provision through joint research

D. Conclusions

We are rather optimistic about the contribution of satellite remote sensing data. There are many initiatives to develop and launch commercial research satellites all over the world.

The problems however, are:

(a) Data price:

- Too expensive to purchase continuously (especially for developing countries);

(b) Data analysis/utilization technology:

- Training, education still not very developed
- New technology: data fusion, GIS.

The important task now is to establish community through these activities in order to work with one another.

V. REMOTE SENSING AND GIS IN VIET NAM

*Tran Manh Tuan**

A. Introduction

In Viet Nam, remote sensing promotion started from 1979 with the implementation of the UNDP/FAO Project VIE/76/014, in the Ministry of Forest, and the UNDP/FAO Project VIE/76/011, in the National Centre of Scientific Research of Viet Nam (NCSR), the predecessor of the National Centre for Science and Technology (NCST).

In this paper, we present an overview of the main institutions in Viet Nam which involve remote sensing and GIS and their projects and results.

B. Main institutions, projects and results

1. Ministry of Agriculture and Rural Development

The Forest Inventory and Planning Institute (FIPI) has the following satellite data:

Landsat MSS of 1975-1980 (whole country);
Landsat TM of 1990 (whole country);
Landsat TM of 1993 (whole country);
SPOT images of 1995-1996 (whole country);
Air photos of 1990-1993 (25 per cent of country territory).

Projects of FIPI, and their results, are wide ranging:

(a) The project VIE/76/014 on forest inventory of Viet Nam: a set of national forest maps at scale of 1:1,000,000 and provincial maps at scale of 1:500,000 (Landsat MSS);

(b) The forest inventory project 1983-1985/Viet Nam-Sweden forestry cooperation programme in raw material area of paper mill (RMA): forest maps at scale of 1:50,000 for three provinces (Vinh Phu, Yen Bai, Tuyen Quang) (air photos);

(c) The watershed planning project/Mekong programme/(central highlands): forest and ecological maps at scale of 1:100,000 (SPOT images);

(d) Land use planning project/Viet Nam-Sweden forestry cooperation programme in 1990-1995: land-use and land use planning maps (Landsat TM);

(e) The forest resources assessment 1990 project of FAO/Government cooperative programme (Landsat TM from three collected study areas of Viet Nam);

(f) Natural habitat monitoring in Remote Sensing Application/WWF 1991: forest vegetation map of natural protection areas at scale of 1:100,000 (Landsat TM);

(g) Forest cover monitoring project of Mekong basin, Interim Mekong Committee, 1992-1995;

(h) Rehabilitation of mangrove forest project (RMFP)/Viet Nam-EC cooperation, 1996-1998.

At the National Institute of Agricultural Planning and Projection (NIAPP), the satellite data consist of five to six tapes (CCT) in Landsat TM form of 92 regions in the central highlands (Tay Nguyen) and the south-eastern part, plus Hanoi, Thai Nguyen and other north-eastern parts.

Projects include the following (including the results of the first two):

(a) Project with the Secretariat of the Interim Mekong Committee: changing identification in Srepok;

* National Centre for Science and Technology (NCST), Hanoi, Viet Nam.

(b) Project with the Republic of Korea: changing analysis with GIS application in Yasup region;

(c) Application of GIS for land use optimization in mountainous barren land (project with Canada);

(d) National project for agricultural land extension;

(e) Application of GIS for planning in the national project of Tay Nguyen and the coastal zone in the central part of Viet Nam.

The overall results of NIAPP projects comprise databases and many maps:

(a) Establishment of national database at a scale of 1:1,000,000 for whole country and 1:250,000 for some economic regions;

(b) Establishment of different thematic maps in digital form, ARC/INFO format, such as land-use maps, soil maps, land classification and analysis maps and geomorphologic maps, among others;

(c) Digitization of topographic maps at a scale of 1:50,000 for whole country.

The digital database (UTM system) belonging to the GIS project includes the following maps:

(a) Country map (for whole country); scale 1:1,000,000:

Soil map;
Temperature map;
Rainfall map;
Soil unit map;
Land use planning map;
Geomorphologic map.

(b) Regional level; scale 1:250,000:

Nine ecology maps in Viet Nam;
Soil map for nine ecological maps;
Current land-use map.

(c) Other thematic maps:

Ecological map of Red River basin;
Agro-ecological zoning map of Tay Nguyen (LAEZ);
Ecological map of Mekong River basin (agro-ecological zoning).

2. General Department of Land Administration

There are four main units in the General Department of Land Administration: Remote Sensing Centre, Centre of Science and Technology, Land Inventory and Planning Institute, and Map Publishing House.

Satellite data in the Department include 55 SPOT images, which fully cover the Viet Nam coast. These images were bought in 1995 through the project of evaluation of coast sensitivity in the event of oil spread (taken in the years 1992-1994). Also, in 1997, 140 SPOT panchromatic images were bought, which cover nearly the whole country, for the purpose of topographic mapping (taken in 1995-1996). The next year should have a series of colour airborne data (scale: 1:50,000) for establishment of land-use maps and topographic maps. Digital data and other databases should follow the national standard for public utilization.

The Department has undertaken a number of projects:

(a) Projects for remote sensing centre development: US$ 3.5 million;

(b) Viet Nam-Sweden cooperation in two projects:

- Land management mapping for updating the land data for whole country (US$ 2 million);
- Establishment of the database for land information (US$ 1.5 million) in Phase 1.

(c) The project in negotiation with Norway: technology for sea bottom survey and mapping. This project also aims to establish a marine geographic database.

The results of the Department's work are as follows:

(a) Digital images (SPOT satellite imagery) for whole country in the Remote Sensing Centre;

(b) Airborne images: cover about 90 per cent of Red River Delta and Mekong Delta (dated from 1995 up to present), both digital and analog. In the near future should have data of Quang Nam-Da Nang and Lai Chau;

(c) Map products: 50 per cent of the maps have been done based on latest information. The Remote Sensing Centre should complete the last in the year of 2001 at a scale of 1:50,000. The large-scale map should be done only for some industrial centres;

(d) Digital national atlas has been made by MapInfo software for public users;

(e) Land information database establishment in form of GIS for many users. So far, 20 per cent of map and land information is in digital form;

(f) 'Registration for land utilization is in digital form mostly for southern part; before long it will be done for the whole country;

(g) One other very important product for the government: establishment of GIS for boundary management. It was completed in three provinces (Ben Tre, Ninh Thuan and Thai Binh) in 1996 and should be done in 10 more provinces in 1997. Another very important product is the digital map of country boundaries and changes over time.

3. National Centre for Science and Technology (NCST)

Among the 17 institutes of NCST, the following nine institute are involved in remote sensing and GIS applications:

- Institute of Physics and a branch in Ho Chi Minh City
- Institute of Geography and a branch in Ho Chi Minh City
- Institute of Oceanography and a branch in Haiphong
- Institute of Geology
- Institute of Geophysics
- Institute of Mechanics
- Institute of Information Technology
- Institute of Ecology and Biological Resources
- Institute of Tropical Biology

The projects of these institutes include:

(a) Project VIE/076/011 on remote sensing;

(b) International programme (Intercosmos) on multispectral images;

(c) Research projects with universities and institutions in France, Canada, Europe, Japan and Australia;

(d) Land-use and land-cover maps;

(e) Integrated survey of the Mekong Delta;

(f) Flood control planning of the Mekong Delta;

(g) Integrated survey of Ca Mau Peninsula and Long Xuyen Quaternary period;

(h) Coastal zone management;

(i) Red River Delta environment information system.

Organizational results can be seen in the many regional and international seminars and training courses on remote sensing and GIS. Many thematic maps at country level and regional level – in geography, geology and oceanography – have been produced.

NCST has also developed software:

(a) MAPSCAN. Software to establish map vectorization from scanner, including text identification;

(b) POPMAP. Distributed in over 100 countries by the Population Programme of the United Nations;

(c) PCIA. Image processing software;

(d) WINGIS. The original version was developed by the branch of the Institute of Physics in Ho Chi Minh City, and now is commercialized by Dolsoft Company.

With regard to computer networks, NCST has developed VAREnet and Netnam, the main pioneer of the Internet in Viet Nam. The members of VAREnet and Netnam are R and D institutes, universities, organizations from the government, NGOs, development assistance organizations, and individuals. The main hubs of VAREnet and Netnam are in Hanoi and Ho Chi Minh City; they are connected to the campus networks of Hanoi National University, Hanoi University of Technology, Ho Chi Minh City National University, AIT Centre in Viet Nam, Institut Francophone Informatique and others.

The Web server address is <http://www.ncst.ac.vn>. To see GIS and remote sensing tutorials and document on the intranet, one may type <http://www.ncst.ac.vn/E_docs.htm>.

4. Other institutions

The Ministry of Science, Technology and Environment (MOSTE) is involved in the national GIS project for natural resource management and environment monitoring. The project was started in 1995. The objective of the project is to set up a national GIS serving natural resource management and environment monitoring. The implementation of this project is done through the subprojects, which are managed by the relevant sectors and provinces.

The time of implementation is 1995-1999. The financial resources comprise an annual allocation from the budget of the National Information Technology Programme (in 1995: VND 4.7 b.; 1996: 4.7 b.; 1997: 2 b.). The unit of implementation is the project office, located at the Ministry of Science, Technology and Environment.

The main activities are to:

(a) Prepare an initial design of the national GIS database for natural resource management and environment supervision (1994-1995);

(b) Build up the data regulations on GIS database for natural resource management and environment monitoring;

(c) Implement the subprojects in the provinces and the sectors: nine sectors and 18 provinces in 1995 and 1996. The subprojects of the sectors are not commonly designed with specific information for each sector. However, the subprojects of provinces follow the same regulations of the database, including the common parts for all the provinces, and other specific parts which correspond to the natural resources of the province. The total number of map information layers of each province is 15;

(d) Test the database of the subprojects, concentrating at the project office;

(e) Integrate the database information layers of the subprojects.

The National Environment Agency (NEA) Database Section started applying GIS in 1996; its main task is to establish GIS database for environment monitoring. NEA is part of MOSTE.

In the Ministry of Industry, the General Department or Geology, Geologic Information Institute, is responsible for geologic map digitization (1:500,000, 1:200,000), stored information for 4,000 mineral resource areas and other geologic survey results.

In the General Department of Meteorology and Hydrography, the Sea Meteorological and Hydrography Centre has been working on elevation map digitization (1:25,000) of the coastal area, and has set up a digital elevation model (DEM). National land-use status map digitization (1:250,000) is proceeding, as is dike system map digitization (1:25,000).

Other institutions include the Geographic Technology and Information Centre, Ho Chi Minh City Polytechnic (training in and transferring GIS technology), and Can Tho University, Faculty of Soil Sciences (remote sensing and GIS for research and training).

C. Future main directions of remote sensing and GIS in Viet Nam

It is important that the country (a) exploit the Internet remote sensing and GIS information, especially for training on the Internet, (b) apply new techniques in remote sensing and GIS and (c) establish national GIS for natural resource management and environment monitoring on the Internet.

PART THREE

PAPERS: TROPICAL ECOSYSTEMS

VI. DEPLETION OF MANGROVE FOREST AND ITS CAUSES IN THAILAND

*Darasri Dowreang**

"Mangroves" may be defined as a community of trees or individual trees occupying the intertidal zone of estuaries in tropical areas. Mangrove associations are distributed vertically according to depth and period of inundation. They also support a complex ecosystem of terrestrial and aquatic flora and fauna.

Mangrove forests are important in many ways. Socio-economically, they are a source for production of wood and charcoal. They provide a breeding ground and nursery of terrestrial and aquatic fauna for fisheries, so they have high potential for fishery industry. They are also used for fish and shrimp farming. As for their agricultural uses, coconut and palm oil plantations can be established on the reclaimed mangrove land. They also play an important role in coastal protection, such as protection against tropical cyclones and protection against coastal erosion, as well as forming a trap for sediments.

Mangroves, however, are sensitive to changes in:

- Sediment load
- Salinity
- Water movement
- Over-cutting
- Land conversion
- Nearby land use

Such changes have impacts on mangroves by:

- Reducing productivity
- Reducing fishery production
- Reducing forest products
- Increasing coastal erosion
- Reducing nutrient-rich sediments
- Increasing marine sediment load

Table 1. Changes in mangroves forest area between 1961 and 1996

Year	Mangrove area (sq km)	Depletion (sq km)	Percentage depletion
1961	3 679	–	–
1975	3 127	552	15.0
1979	2 873	254	8.1
1986	1 964	909	31.6
1991	1 736	228	11.6
1993	1 687	49	2.8
1996	1 676	11	0.7

Source: Royal Forest Department.

* National Research Council of Thailand, Bangkok.

Table 2. Land-use types in the area designated as mangrove zones (from 1996 Landsat images)

Land-use types	Area (sq km)					Percentage
	Eastern	Central	Southern (Gulf)	Southern (Andaman)	Total	
1. Mangroves	126.58	54.49	165.71	1 329.04	1 675.82	45.0
2. Shrimp farms	242.95	156.29	219.20	51.54	669.98	18.0
3. Settlements	39.57	30.99	10.01	7.42	88.01	2.4
4. Others	139.35	428.04	169.57	553.72	1 290.67	34.6
Total area	548.54	669.81	564.49	1 941.72	3 724.48	100.0

Source: Royal Forest Department.

Mangrove depletion has several serious causes. Over cutting is the main cause. Concessions of 1,695 sq km of forest area yield approximately 952,845 cu m/year. The capacity of charcoal burners is approximately 1,548,682 cu m year, so illegal cutting no doubt is rampant. Forest encroachment is caused by expansion of agricultural land and resettlements in the forest resulting from new developments such as construction of roads through the forest. Shrimp farming accounts for about 670 sq km (18 per cent) of mangrove zonation. About 9.3 sq km of mangroves area have been used for mining activities. And the establishment of industrial estates along the coast, as well as seafood product factories within the forest, have also contributed to mangrove depletion.

New measures for mangrove forest management are being tried. Management is taking place at provincial level (22 provinces) in three regions: Central (5), Eastern (5) and Southern (12). Also, the government has come up with four strategic plans for 1999-2003:

- Building up public awareness and concern in the area of environment;
- Protection and prevention;
- Rehabilitation/reafforestation;
- Practical research (e.g. in biodiversity, reutilization of shrimp farms and other areas).

Bibliography

Charuppat, T. and J., 1997. *Application of Landsat-5 (TM) for Monitoring the Changes of Mangrove Forest Area in Thailand.* Project No. C-09-07-72-40-001, Royal Forest Department.

Hodson, William M., 1983. *Mangrove Biological Processes and the Role of Remote Sensing.* Report of Regional Remote Sensing Training Course of Mangrove Ecosystem, Bangkok, 28 November - 16 December 1983.

Wacharakitti, S., 1983. *Mangrove Ecosystem: In General.* Report of Regional Remote Sensing Training Course of Mangrove Ecosystem, Bangkok, 28 November - 16 December 1983.

VII. FOREST MAPPING IN VIET NAM

*Nguyen Manh Cuong**

Introduction

In any programme or project concerning natural or environmental inventory, maps are always an important source of data, particularly in forestry. Because forests are spread out over large and remote areas and are constantly being changed by many different causes, forest mapping is difficult work.

For many years, the Ministry of Forestry of Viet Nam has paid much attention to forest inventory, and forest mapping in particular. Since 1960, aerial photos have been used for forest surveys and mapping. But before 1979, using aerial photos and traditional methods, forest maps in Viet Nam had been established for the micro-level at enterprise, district and province levels only. Starting in 1979, with support from the Food and Agriculture Organization of the United Nations (FAO), and using remote sensing techniques for the first time, forest maps for the whole territory of Viet Nam were established over a two-year period. This was an important event in the history of forest mapping in Viet Nam. Remote sensing techniques, using satellite images, had opened the door wide to improvement of the capability of forest mapping in Viet Nam.

A. An introduction to forest inventory and remote sensing applications for forest assessment and monitoring in Viet Nam

1. Forest assessment and monitoring

In previous decades, forest inventories were carried out at two levels (macro for national or provincial master planning and micro for enterprise management planning), with emphasis given only to rich forests to help logging planning rather than forest management and protection.

No computerized databases or permanent sample systems for long-term monitoring of forest resources were made. The National Forest Inventory Project (1981-1983), supported by the United Nations Development Programme (UNDP), used remote sensing data (Landsat MSS images) for forest mapping and temporal sample plots for volume estimation, but a long-term monitoring basis was not mentioned. Therefore, most of the data and information were inadequate, inaccurate and often not even available for use; in many cases, they were not homogeneous in their dates, classification system or technical procedures for collection, processing and statistics. These shortcomings have limited the data comparison and integration that needs to be carried out.

According to inventory data of 1943, Viet Nam once had a large forest area that covered 42 per cent of the country (about 14,000,000 ha), but after 40 years, the forest area had decreased greatly, and by 1990, Viet Nam had only 9.2 million hectares of forest as shown in table 1 below.

2. Forest Resource Assessment and Monitoring Programme 1991-1995

To protect valuable forest resources as well as the living environment, the Government of Viet Nam has paid a great deal of attention to multiple forest resource inventory (MFRI) as well as sustainable forest management.

*Remote Sensing Section, Forest Inventory and Planning Institute (FIPI), Hanoi, Viet Nam.

Table 1. Forest cover in Viet Nam, 1990

Type	Area
Natural forests	8.5 million hectares
Woody forests	6.8 million hectares
Bamboo forests	1.1 million hectares
Mixed forests	0.6 million hectares
Plantation forests	0.7 million hectares

Source: Forest Inventory and Planning Institute.

On 27 November 1993, at the request of the Ministry of Forestry (MOF), the Prime Minister of Viet Nam officially approved the national programme, Forest Resources Assessment and Monitoring Programme for the period of 1991-1995 (FRAMP 1991-1995) (see figure I). The Forest Inventory and Planning Institute (FIPI) is under the general jurisdiction of MOF (now known as the Ministry of Agriculture and Rural Development (MARD)) and has responsibility for state management in the field of forest inventory and planning. FIPI is organized as a hierarchical system, from FIPI Headquarters to sub-FIPIs located in different regions of the country, to carry out its functions and duties.

FIPI has successfully completed FRAMP 1991-1995 and is carrying out FRAMP 1996-2000.

Being aware of the usefulness of the new remote sensing and geographic information system (GIS) technology, FIPI has applied them as an integrated method for MFRI in FRAMP 1991-1995, and plans to continue them at a higher level in FRAMP 1996-2000.

(a) Objectives of FRAMP 1991-1995

Since 1991, in parallel with the further strengthening of the forest management system, FIPI has conducted the first continuous FRAMP with the following objectives:

(a) Assessment of forest resources as seen from the broad concept of a forest ecosystem. From this assessment, highly accurate data and information will be supplied annually and periodically for both forest management planning and long-term forest monitoring at the macro level;

(b) Establishment of a computerized database and GIS system and a permanent sample system as a basis for long-term monitoring and research study of forest resources;

(c) Assessment of the changes in forest resources during the periods 1976-1990 and 1990-1995, which will help in forest forecasting for the next periods.

The main contents of the chart of FRAMP 1991-1995 include the following:

Permanent sample system and field inventory method;
Forest mapping method;
GIS use;
Data processing method and establishment of a data bank;
Study on forest change.

(b) Permanent sample system and field inventory method

A two-stage sample system was designed for the permanent sample system and field inventory method. Some 3,000 primary units (PU), with 100 hectares each (1 km x 1 km in size) (see figure II), are systematically distributed in a grid of 8 km x 8 km covering the whole forested area, and one fifth of them are taken every year for data collection, which consists of land-use type mapping for the whole PU, and forest measure and other observations in 20,000 sq m of an L-shaped secondary unit (SU) located in the centre of the PU (see figure III).

The following information and data will be recorded in both PU and SU for subsequent analysis:

Land use mapping in PU, with emphasis on forests, roads and residential lands;

Forest measure, including species composition, DBH (diameter-base-height), height, tree quality, species, forest and soil profiles, non-wood forest products;

Pests and diseases, wildlife and insects;

Demographic information of local communities, including population, ethnic group, land productivity, income and other information related to their dependence on forests.

(c) *Forest mapping method*

The forest maps are established using remote sensing data (mainly Landsat TM) combined with aerial photos (where available), other available data and field checks. This was done twice during FRAMP, the first in 1991 and the second in 1995, to enable a two-date comparison. In the FRAMP 1991-1995 period, forest maps of 1992, 1993 and 1994 were established as updated maps. Forest maps are produced at scales of 1:1,000,000 for the whole country, 1:250,000 for each forestry region and 1:100,000 for provincial land-form maps. Maps will be digitized and stored in computer by using GIS techniques for area measurement and production of thematic maps (see figure I).

(d) *GIS use*

The GIS use in FRAMP 1991-1995 was the first application of GIS to forestry in Viet Nam. Because of many limitations, such as technology, equipment and professional experience, primary results of the GIS use for MFRI data presentation and analysis are still limited.

The map processing in FIPI is done mainly by using a PC with simple software such as AutoCAD and MapInfo. Some other standard GIS software, like ARC/INFO, SPANS, ILWIS, PANMAP and IDRISI, have been installed, but very few can carry out the large volume of map work in FRAMP 1991-1995. On the hardware side, no workstations or peripheral equipment are available.

Through great effort, FIPI digitized all of the final maps of FRAMP 1991-1995 (including the forest maps of 1991, 1992, 1993, 1994 and 1995 at a scale of 1:250,000 for each region and for the whole country at a scale of 1:1,000,000 as well as the land-form maps at 1:100,000 for provinces), using MapInfo software for map editing and data management. The following information layers have been stored in databases:

Forest and land-use layer for each year;

Contour lines of 100, 200, 300, 500, 700, 1,000, 1,300, 1,500, 1,700, 2,000, 2,300, 2,500, 2,700, 3,000 and 3,300 m;

Layer of soil type;

Layer of boundary system of forest management units, such as enterprises, protected areas and others;

Layer of administrative boundary system from province up to commune;

Layer of transport and hydrological systems;

Layer of local names;

Layer of national geographical coordination.

Using these information layers, a number of thematic maps of different territory units are made by also incorporating GIS:

Slope map;
Map of height class;
Forest ecological map;

Land-form map;
Land-site map;
Watershed planning map;
Land use planning map;
Land use suitability map.

Through MFRI data analysis, the primary results gave support to the concept of forest management, as did studies on forest resource change monitoring.

(e) Data processing method and establishment of a data bank

Within the framework of FRAMP 1991-1995, as described above, data processing and the establishment of a data bank were completed by the integration of GIS and field inventory data. The data from GIS and field inventory data, particularly the data from 3,000 permanent plots, are important resources for estimating forest resource data. Three main factors reflected in the content of forest resource data are S (area), M (average volume per ha) and DBH for each forest type.

The data processing procedure is shown in figure IV.

(f) Main results of FRAMP 1991-1995 in database and forest change study

Table 2. General data on the forests in Viet Nam in 1995

Areas \ Land types	Total area	Forest cover			Non-forest land	Other land
		Total	Natural forest	Land plantation		
Whole country	33 111.6	9 302.2	8 252.5	1 049.7	9 778.6	14 030.8
North-West	3 595.3	515.5	464.1	51.4	2 232.0	847.8
North-Centre	3 332.5	806.7	667.2	139.5	1 441.1	1 084.7
North-East	3 368.8	670.7	533.8	136.9	1 388.6	1 309.5
North Delta	1 251.2	53.4	22.7	30.7	37.7	1 160.1
North Coastal	5 118.3	1 791.9	1 564.6	227.3	1 491.6	1 834.8
South Coastal	4 587.5	1 597.4	1 439.8	157.6	1 331.9	1 658.2
Plateaus	5 556.8	3 168.1	3 108.9	59.2	1 264.7	1 124.0
South-East	2 345.0	486.2	406.8	79.4	361.9	1 496.9
Mekong Delta	3 955.7	211.8	44.6	167.2	229.1	3 514.8

Table 3. Forest cover (in per cent) of the country and each region

Areas	Forest cover (percentage)
Whole country	28
North-West	14
North-Centre	24
North-East	20
North Delta	4
North Coastal	35
South Coastal	35
Plateaus	57
South-East	21
Mekong Delta	5

(g) Study of forest change

Study of forest change is one of the important areas under FRAMP 1991-1995. Case studies on deforestation use available data combined with data newly collected from some test sites.

Table 4. Data in functional forest areas

Unit: 1,000 ha

Forest types areas		Total	Production	Protection	Special use
Whole country	Forest	9 302.2	4 925.2	3 478.7	898.3
	Non-forest	9 778.6	5 701.3	3 269.5	807.8
	Total	19 080.8	10 626.5	6 748.2	1 706.1
North-West	Forest	515.5	134.0	319.3	62.2
	Non-forest	2 232.0	863.1	1 034.7	334.2
	Total	2 747.5	997.1	1 354.0	396.4
North-Centre	Forest	806.7	345.8	414.9	46.0
	Non-forest	1 441.1	932.2	463.1	45.8
	Total	2 247.8	1 278.0	878.0	91.8
North-East	Forest	670.7	423.9	216.4	30.4
	Non-forest	1 388.6	871.6	367.9	149.1
	Total	2 059.3	1 295.5	584.3	179.5
North Delta	Forest	53.4	18.3	10.3	24.8
	Non-forest	37.7	18.4	8.0	11.3
	Total	91.1	36.7	18.3	36.1
North Coastal	Forest	1 792.4	862.1	723.7	206.6
	Non-forest	1 491.6	812.7	617.6	61.3
	Total	3 284.0	1 674.8	1 341.3	267.9
South Coastal	Forest	1 597.4	798.6	728.2	81.4
	Non-forest	1 331.9	1 201.1	472.9	60.4
	Total	2 929.3	1 999.7	1 201.1	141.8
Plateaus	Forest	3 168.1	2 016.8	810.3	341.0
	Non-forest	1 264.7	999.2	169.1	96.4
	Total	4 432.8	3 016.0	979.4	437.4
South-East	Forest	486.2	223.8	180.6	81.8
	Non-forest	361.9	229.0	109.6	23.3
	Total	848.1	452.8	290.2	105.1
Mekong Delta	Forest	211.8	112.7	75.0	24.1
	Non-forest	229.1	176.5	26.6	26.0
	Total	440.9	289.2	101.6	50.1

However, like many agriculture-based countries in the region, Viet Nam has been suffering for a long time from heavy and seemingly endless deforestation caused by different factors (both natural and man-made) that may simultaneously strike forests, leading to numerous difficulties for the quantitative assessment of forest changes and for identifying key factors causing forest loss. The following activities have been performed for the purpose:

(a) Collection of available data, including forest and land-use maps, reports and figures, forming a series of data corresponding to each study time-mark such as 1976, 1990 and 1995. Multi-date comparisons are made to identify differences, which are analysed according to their relationship with different socio-economic and natural conditions, thus, it is hoped, providing reasonable explanations in each case;

(b) Selection of test sites, based on the following criteria:

(i) To be representative of deforestation;

(ii) To have as many factors as possible causing forest loss.

FAO procedures for assessment of forest change were adapted for the work on Viet Nam's conditions in the following ways:

(a) Forest maps were established using remote sensing data and aerial photos. Landsat MSS images were used for the 1976 forest map and Landsat TM for the 1990 forest map. Aerial photos taken during 1983 and 1989 were also used as associated data;

(b) The study area was covered with a sample dot grid; each dot bears a sample plot 2 km x 2 km in size at map scale 1:250,000 at macro level, and 0.5 km x 0.5 km at map scales 1:50,000 to 1:25,000 at micro level;

(c) Plot description was made and the following information was identified: forest type, topographical type, soil type, rainfall, humidity, population, ethnic group, road density, management status and others. These will be stored in computers and used for regression analysis to assess forest change corresponding to each specific condition;

(d) FoxPro software was used for calculating forest resource change rate and establishing the matrix of its change, as well as for modelling the relation between forest cover and population density.

The work was implemented in this way for three out of four study regions selected as representative of forest change in the whole country.

Tables 5 and 6, based on multi-data analysis at country level, show forest area change in Viet Nam in the last 20 years.

Table 5. Forest area change in 20 years

Unit: 1,000 ha

	1976	1980	1985	1990	1995
Forest cover	11 169.3	10 608.3	9 891.9	9 175.6	9 302.2
Natural forest	11 076.7	10 186.0	9 308.3	8 430.7	8 252.5
Forest plantation	92.6	422.3	583.6	744.9	1 047.7

Forest cover percentage (compared to total country area) has been decreasing over the years, as shown below.

Table 6. Decreasing forest cover

Year	Forest cover (percentage)
1943	43.2
1976	33.7
1990	27.7
1995	28.1

3. Remote sensing applications for forest mapping

Remote sensing applications for forest mapping and monitoring have been carried out by the Remote Sensing Section of the Forest Inventory and Planning Institute. The Section was established in 1962 for aerial photo applications in forestry. Since 1980, remote sensing data and technology have been used to establish thematic maps with suitable contents for the country, regions and provinces, even districts and specific study areas, at different scales. The Section has experience in applications of different types of satellite data, such as Landsat MSS, Landsat TM, SPOT and MKF-6. The work has been carried out by visual interpretation and digital image processing in integration with GIS.

(a) Completed projects in Viet Nam using remote sensing data

(a) Project VIE/76/014 on forest inventory of Viet Nam: a set of national forest maps at a scale of 1:1,000,000 and provincial maps at a scale of 1:500,000 were established, based on Landsat MSS;

(b) Forest inventory project 1983-1985/Viet Nam – Sweden forestry cooperation programme in raw material area of paper mill: present forest maps at 1:50,000 scale for three provinces (Vinh Phu, Yen Bai and Tuyen Quang) established from aerial photos;

(c) Watershed planning project/Mekong programme (Western Plateau and Tay Nguyen) in 1987: thematic maps (forest and ecological maps) at a scale of 1:100,000, based on SPOT images;

(d) Land use planning project/Viet Nam – Sweden forestry cooperation programme in 1990-1995: present land-use and land use planning maps based on Landsat TM are in progress;

(e) Forest resources assessment 1990 project of FAO/Government Cooperative Programme: Landsat TM data from three study areas of Viet Nam were used to establish a database for forest assessment and monitoring;

(f) Natural habitat monitoring in remote sensing applications/World Wildlife Federation, 1991: forest vegetation map of natural protection areas and national parks (Tam Dao, Muong He, Nam Cat Tien and others), based on Landsat TM, scale 1:100,000;

(g) Forest cover monitoring project of Mekong basin – Interim Mekong Committee, 1992-1995: the forest cover map of the Mekong basin area in different periods was established;

(h) Rehabilitation of mangrove forest project/Viet Nam – Euroconsult cooperation, 1996-1998;

(i) Information system development for tropical forest management/FIPI – Japan Forest Technical Association (JAFTA) cooperation, 1997-1998: forest maps for six of seven important forest regions will be established based on remote sensing techniques (north-western, north-central, north-eastern, south coastal, highland and south-eastern regions);

(j) Study on development of technology for rehabilitation of devastated tropical forest/FIPI-JAFTA cooperation, 1997;

(k) Different research studies and all national projects carried out by FIPI with participation of the Remote Sensing Section;

(l) Job training offered by the Remote Sensing Section on applied remote sensing techniques to the Forest College of Viet Nam and some foreign students, for example from the Lao People's Democratic Republic, Cambodia and Gent University in Belgium.

(b) Completed projects in other countries

(a) The Viet Nam – Lao People's Democratic Republic cooperation programme: forest maps of the Lao People's Democratic Republic, 1982-1990;

(b) The Viet Nam – Cambodia cooperation programme: forest maps of Cambodia, 1982-1984.

4. Technical capacities and data used in the Remote Sensing Section

The Section possesses a relatively well-equipped laboratory, with the following equipment:

Electric-optical equipment for visual interpretation: stereoscopes, stereo-facet plotter, zoom transfer scope, Pantophot additive colour viewer AC-90B, light tables;

Hardware and some peripherals for digital image processing and GIS: 486-PC, 586-PC, HP PenJet XL300-A3 colour printer, and digitizer for A3 and A0 sizes;

Software packages: ILWIS 1.41, ARC/INFO, ArcView, SPAN, MapInfo, IDRISI and others.

The remote sensing data used include:

Landsat MSS of 1975-1980 over the whole country;
Landsat TM of 1990 covering the whole country;
Landsat TM of 1993 for the whole country;
SPOT images of 1995-1996 over the whole country;
Aerial photos of 1990-1993 covering 25 per cent of the country.

B. Method of forest mapping based on remote sensing data

According to the technical introduction of the Forest Inventory and Planning Institute, the main steps of the forest mapping method based on satellite images are as shown in figure V.

1. Material collection and preparation

(a) Remote sensing data used

The satellite images from Landsat MSS, Landsat TM and SPOT are used as the main sources for photo-interpretation. Depending on the objective, the contents of the map and map scales, the most suitable type of satellite images should be chosen for interpretation:

(a) Landsat MSS is suitable for forest mapping at scales from 1:1,000,000 to 1:500,000 (maximum scale 1:250,000);

(b) Landsat TM is suitable for forest mapping at the scale of 1:250,000 (maximum 1:100,000);

(c) SPOT images are suitable for forest mapping at the scale of 1:100,000 to 1:50,000 (maximum 1:25,000);

(d) The remote sensing data used for present forest mapping should not be older than one year from the date of mapping and should be processed in geometric correction;

(e) All kinds of thematic maps concerning the study area, such as historical forest map, land-use map, ecological map, soil map and topomap, may also be used as important reference sources in photo-interpretation. In this step, the aerial photos play a very important role.

(b) Base maps used

Base maps used for forest mapping are topographical maps, which include the following scales and projections:

1:1,000,000 Gauss or geographical projection;
1:250,000 UTM projection;
1:100,000 UTM projection;
1:50,000 UTM projection.

(c) Establishing classification

To classify forest objects, ecology is used as the basis, and factors like forest cover, historical sources, distribution and species are used for the division into subclasses and objects. This point of view makes a full picture of natural distribution of the forest and land-use objects, as well as increasing map information for multi-purpose studies such as ecology, forest resources, land use and so on.

Depending on what kind of satellite and images are chosen, and depending on the map scales and objectives of the map, a suitable classification must be established for photo-interpretation.

In accordance with economic and geographic characters, Viet Nam may be divided into nine regions, as in figure VI.

I. North-West
II. North-Central
III. North-East
IV. Red River Delta
V. North Coastal
VI. South Coastal
VII. Highland Plateau
VIII. South-East
IX. Mekong Delta

As a result of good natural conditions and climate, Viet Nam has a diversity of forest types, with an abundance of species composition. Below are some typical forest types:

Evergreen and semi-evergreen forests;
Deciduous forests;
Mangrove forests;
Melaleuca (or cajeput) forests;
Coniferous forests (mainly pinus);
Pure bamboo and mixed bamboo and wood forests;
Plantation forests: pinus, eucalyptus, mangletia, styrax, rhizophora, casuarina.

Table 7 shows the official classification made by FIPI based on Landsat TM, for the national forest resource assessment and monitoring programme, 1991-1995.

Based on the FAO classification system, the forest cover percentage is as follows:

Close	> 70	per cent
Medium	40-70	per cent
Open	< 40	per cent

2. Establishing photo-interpretation (photo-keys) of the objects distributed in the area

In photo-interpretation, particularly for areas where there is not enough good information for reference and interpretation, the photo-keys play an important role. Photo-keys are understood as a complete set of typical characters for each object or any object group in the satellite image. The photo-key of each object is shown through two typical factor groups:

(a) Direct factor group:

Colour, tone;
Texture, geographical position;
Form, acquired data.

(b) Indirect factor group:

Location of distribution: near or far from industrial area, settlement and so on;

Social and economic character of the study area;

Ecological, climatic, hydrological and other characteristics;

Land-use characteristics of the area (traditional land use);

Government policies on socio-economic development concerned with the land-use situation in the territory.

All of the factors above are combined in setting up photo-keys and supporting the interpretation. The interpreters have to remember those typical factors and analyse them in the interpretation process.

3. Production of a draft map

Production of the draft map involves several steps:

(a) Photo-interpretation;
(b) Photo-transfer to the base map;
(c) Field checking and correction;
(d) Completing the author map;
(e) Editing the final map and calculating the area.

Table 7. Legend system of forest map

Content of the map	For scale of 1:1 000 000	For scale of 1:250 000 and 1:100 000
A. Land surface area		
I. Forest cover land	x	x
1. Natural forest	x	x
1.1 Closed forest type	x	x
1.1.1 Broadleaf forest	x	x
1.1.1.1 Evergreen forest:	x	x
- Close		x
- Medium		x
- Open		x
Mangrove forest:	x	x
- Close		x
- Medium and open		x
Rocky forest	x	x
1.1.1.2 Semi-deciduous forest:	x	x
- Close		x
- Medium and open		x
1.1.1.3 Deciduous forest:	x	x
- Close		x
- Medium and open		x
1.1.2 Coniferous (pine)	x	x
1.1.2.1 Pure pine forest:	x	x
- Close		x
- Medium and open		x
1.1.2.2 Mixed pine forest:	x	x
- Close		x
- Medium and open		x
1.1.3 Bamboo forest	x	x
1.1.3.1 Pure bamboo forest	x	x
1.1.3.2 Mixed bamboo forest	x	x
1.2 Open forest type (Dipterocarpus)	x	x
1.2.1 Broadleaf forest	x	x
1.2.2 Conifer	x	x
2. Forest plantation		x
2.1 Broadleaf forest		x
2.1.1 Evergreen forest		x
2.1.1.1 No mangrove forest		x
2.1.1.2 Mangrove forest		x
2.2 Conifer (pine) forest		x
2.3 Bamboo		x
II. Non-forest land	x	x
1. Non-land-use area	x	x
1.1 Woodland	x	x
1.2 Savanna (shrub, grass)	x	x
1.3 Mosaic	x	x
1.4 Open rocky	x	x
1.5 Sand area	x	x
2. Land use	x	x
2.1 Agriculture land use	x	x
2.1.1 Short-term plantation		x
2.1.2 Long-term plantation		x
2.2 Grassland		x
2.3 Settlement	x	x
2.4 Building		x
2.5 Roads	x	x
B. Water bodies	x	x
I. Lake, river	x	x
II. Swamp	x	x

(a) Photo-interpretation

Table 8. Technical standards in drawing

Map scale	Minimum area to draw	
	In the map	In the field
1:500 000	6 mm	150 ha
1:250 000	6 mm	38 ha
1:100 000	6 mm	6 ha
1:50 000	6 mm	15 ha

In the visual interpretation method of performing photo-interpretation on a colour composite picture enlarged to the scale of the base map, the technical operation is as follows:

(1) Transparency film of the same size is placed over the picture and kept in place by glue transparency band; the minimum area that is draw is six hectares;

(2) Picture code number and date are written on the left corner of the film, and coordination lines are drawn with a black pen;

(3) The hydrological system (rivers, streams, channels and so on) are drawn in the picture in blue ink;

(4) The road system is drawn in red ink;

(5) Step by step, the forest, non-forest, land-use and non-land-use blocks are drawn with black pen (size 0.1-0.25 mm);

(6) Then, step by step, separate objects are drawn inside each block above;

(7) All of the objects are coded by regular numbers.

Some tools used for interpretation are loupe glass, light table and other basic tools.

Checking after interpretation

During the interpretation process some technical mistakes happen, such as contour lines are not closed, there is no code number in the polygon, or some objects are lost, so checking and correction take place after interpretation to correct all of the mistakes and make a good draft map.

(b) Photo-transfer to the base map

All of the results from photo-interpretation must be exactly transferred to the base map. In this case, the SPOT image is enlarged and geometrically corrected at the same scale as the final map, 1:100,000, so that in general the topology factors in the picture fully cover the base map. But in fact, some factors are not covered by topology changes, so some technical operations have to be carried out in the transfer process.

The transfer method can be carried out for two cases: full cover and not full cover.

(i) Full cover case

Technical operation for this case is very simple and includes the following operations:

(1) Take the transparency film off the picture and put the film on a light table;

(2) Put the base map on the transparency film and adjust the base map so that it fully covers the film. Pay attention that roads and the hydrological system are fully covered (error $\leq$ 1 mm for the whole map sheet);

(3) Redraw exactly all of the polygon lines made by interpretation and write code numbers inside polygons;

(4) Using official materials, redraw the boundary system (provinces, districts, communities, enterprises), as well as local names, on the base map.

(ii) Not full cover case

Depending on the changing level between picture and base map, the transfer method for not full cover is applied as follows:

(1) Define points or lines that are not changed in both base map and film, and then adjust base map to cover fully the film;

(2) After doing that, proceed as in the full cover case. The lines and objects that are not covered are changes or new events (alluvial or erosion areas, new channels, roads and so forth).

(iii) Checking results of the transfer

Checking focuses on technical mistakes, as mentioned above.

(c) Field checking and correction

(i) Content and request of the field checking

The technical standards of the interpretation, map scale, legend system and transfer compose the contents and request of the field checking. For each object, the field working group must define and come to a conclusion on the following:

Geographic location and terrace characteristics of the base map;
The object's name, as mentioned in the legend system;
Object boundaries, particularly for new changes.

(ii) Method

Field checking and correction are carried out as follows:

(1) Checking and correction (work indoors): using new information from the forest inventory map made by local organizations, check and correct all areas that are not clearly interpreted or that have changed (since the satellite image used may be old). This result is the first correction;

(2) Field checking and correction by tours and points:

(a) Checking tours are defined to pass most of the important places in the area;

(b) Control points are defined for different forest types in the map, particularly for objects that are not clear in the picture, such as new forest plantations, scattered regrowth forest, open forest in shrimp farms, and others; such objects are carefully checked and corrected.

(d) Completing the author map

After the field work, the map is checked again for technical mistakes and corrected to obtain the following technical standards:

The correct code number is clearly written inside each object;

Object boundaries are closed;

All boundaries of administrative units, enterprises, protected coastal areas and so on are complete;

The systems of roads, rivers, streams and channels are shown as in the base map used;

The legend table is clearly shown in a corner of the map.

After completing the checks and corrections, the map is called the "author map" and is sent for editing of the final map ("output map").

(e) Editing the final map and calculating the area

(i) Editing the final map

To carry out the full mapping process, the final map is edited by one of the two following methods:

(a) *Traditional method.* The final map is edited (redrawing and putting in colours) by hand on the projection map (contour lines and some topology factors are taken off);

(b) *Modern method.* Geographic information system is applied, based on computer operation and professional software.

(ii) Area calculation method

The area calculation can be carried out by a grid system (or from a data table file in the computer, after digitizing and implementing the polygon file). The grid system is as follows:

(1) On millimetre-grid paper, make a point in black ink at central points of 4 sq mm. Thus, each black point will represent an area of 4 sq mm in the map at 1:100,000 scale (equal to four hectares in the field);

(2) Cover the author map with millimetre paper and put them on the light table;

(3) Count the black points distributed inside each polygon and write the total number of points in the statistics list (a point located on the boundary of two polygons has a value of 0.5 point);

(4) The area of each object (in hectares) is written for that object;

(5) Multiply the total number of points in the map by four to get the total size of the project area (see figure VII).

Bibliography

Cuong, N.M., 1991a. *Album of Photo-keys for Photo-interpretation of Landsat TM* (Hanoi, Forest Information and Planning Institute (FIPI)).

_______, 1991b. *Forest Mapping Method at Small Scale Based on Remote Sensing Data* (Hanoi, FIPI).

Food and Agriculture Organization of the United Nations, 1989. *Classification and Mapping of Vegetation Types in Tropical Asia* (Rome, FAO).

Forest Inventory and Planning Institute, 1996. *Report on Forest Resource Assessment and Monitoring Programme 1991-1995* (Hanoi, FIPI).

Sing, K.D., 1987. The application of SPOT images for forest resources assessment and monitoring report on a meeting on the application of SPOT images (Singapore).

Sukhik, V.I., 1979. *Remote Sensing Methods in Natural Protection and Forestry* (Moscow, Publishing House of Forest Industry).

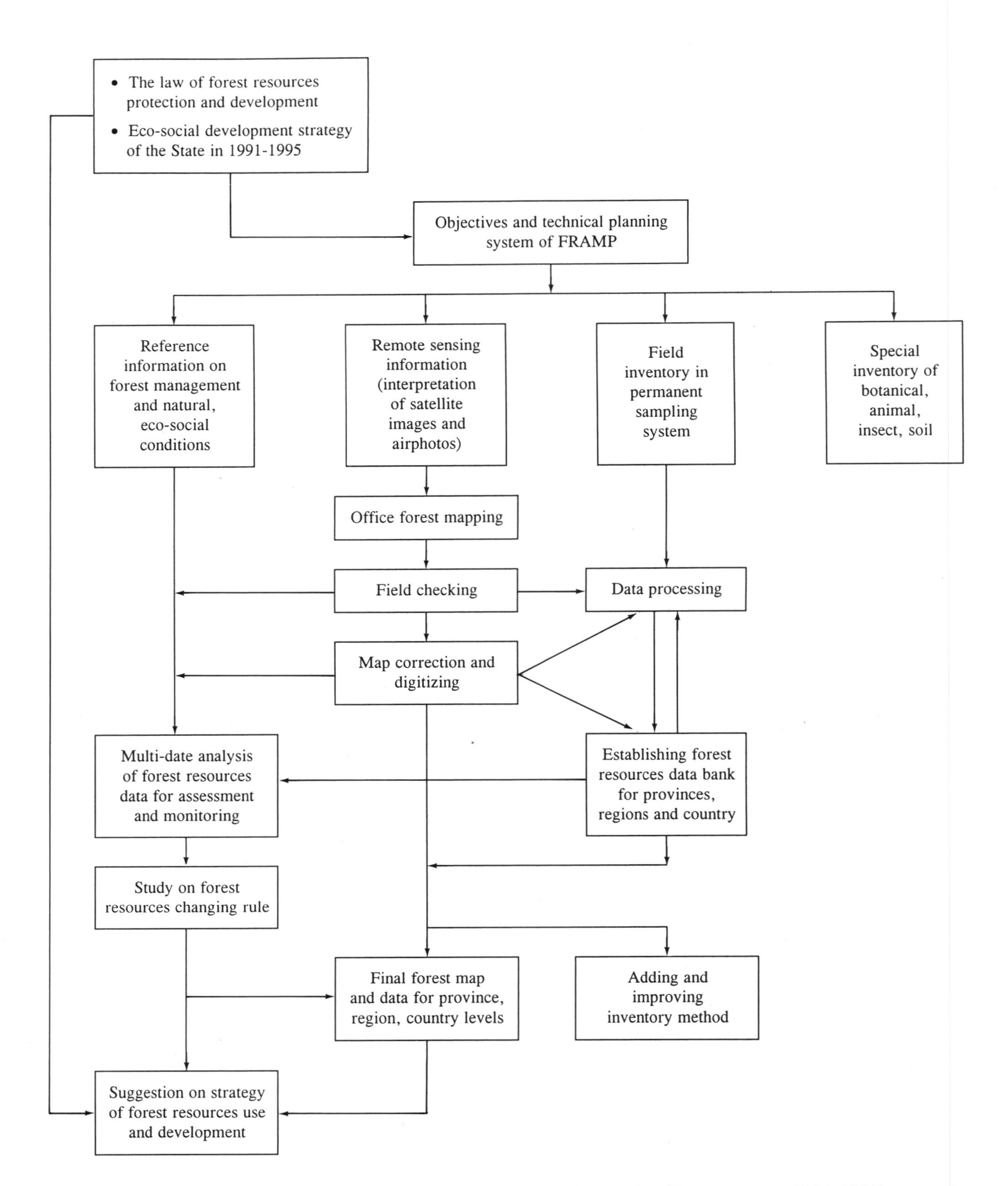

Figure I. Chart of the Forest Resource Assessment and Monitoring Programme, 1991-1995

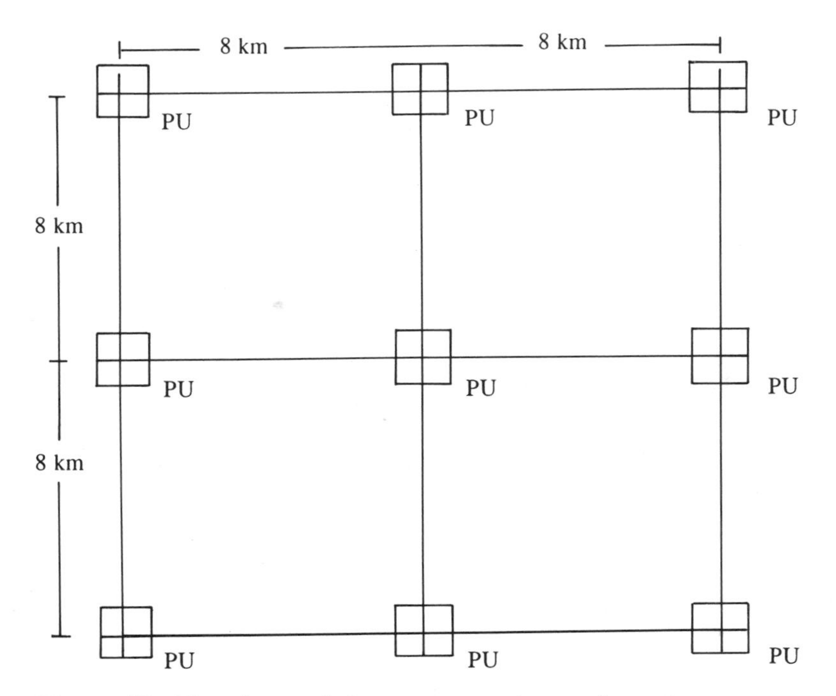

Figure II. The chart of the permanent sample system

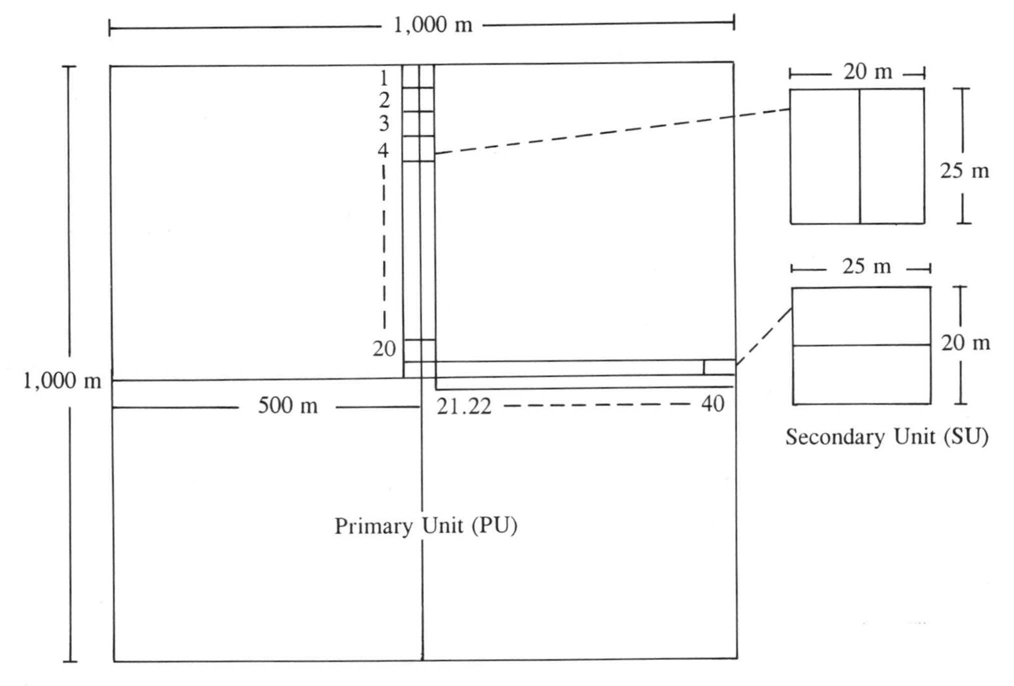

Figure III. The chart of primary and secondary units

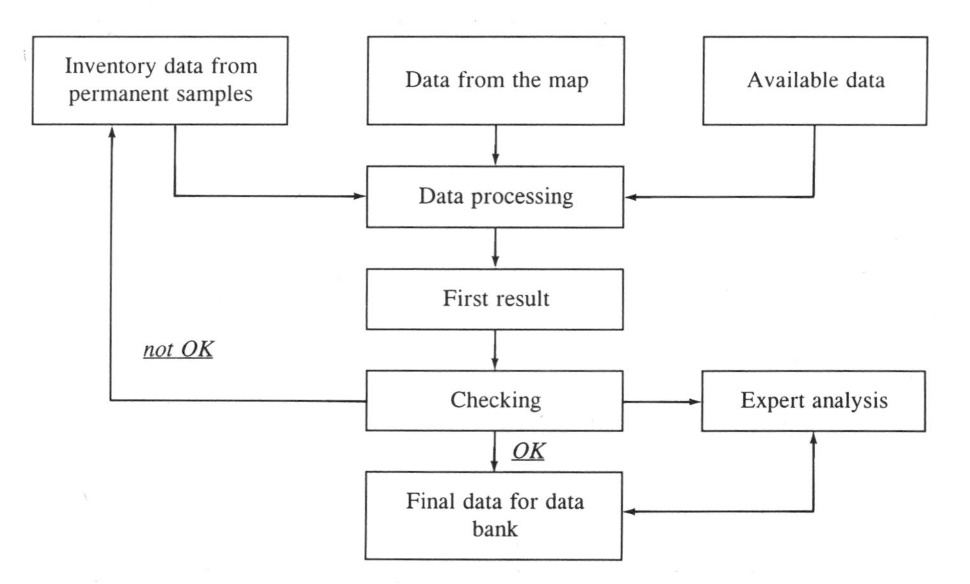

Figure IV. Data processing procedure

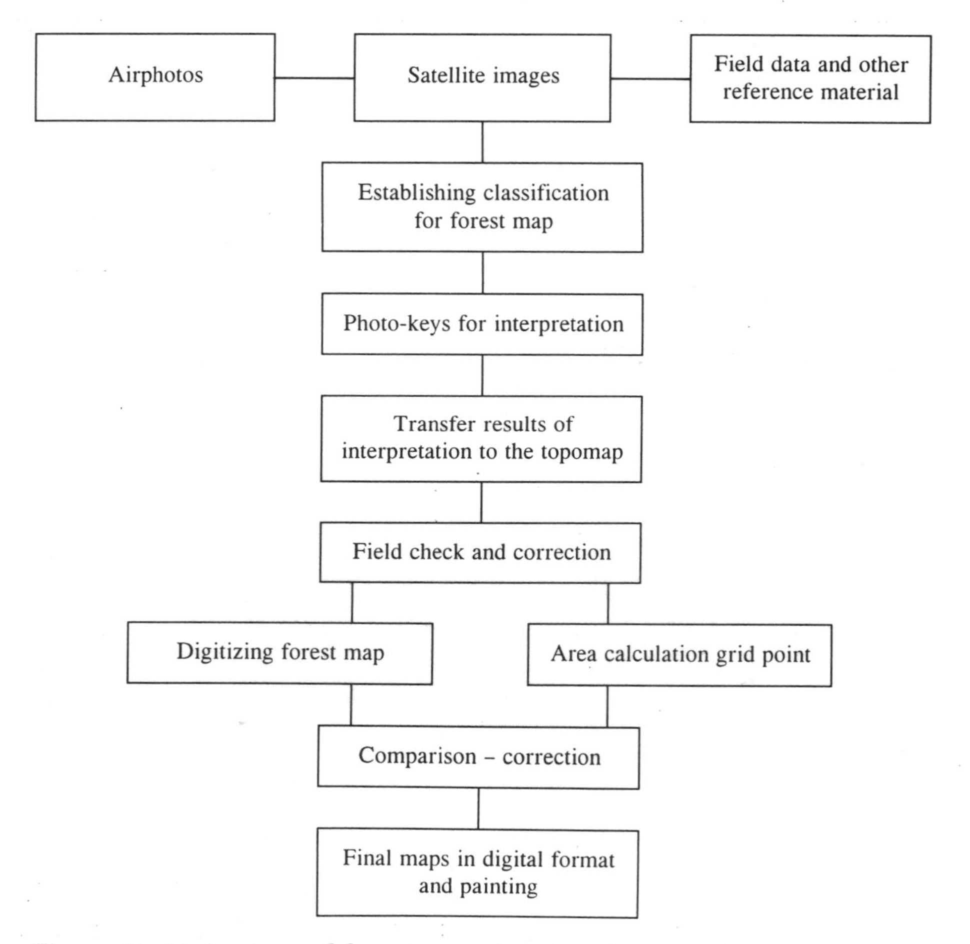

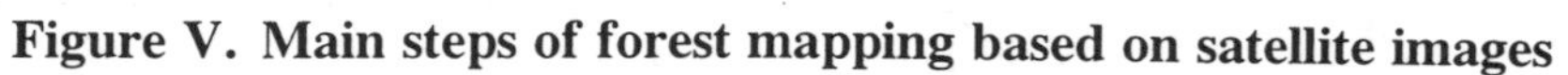

Figure V. Main steps of forest mapping based on satellite images

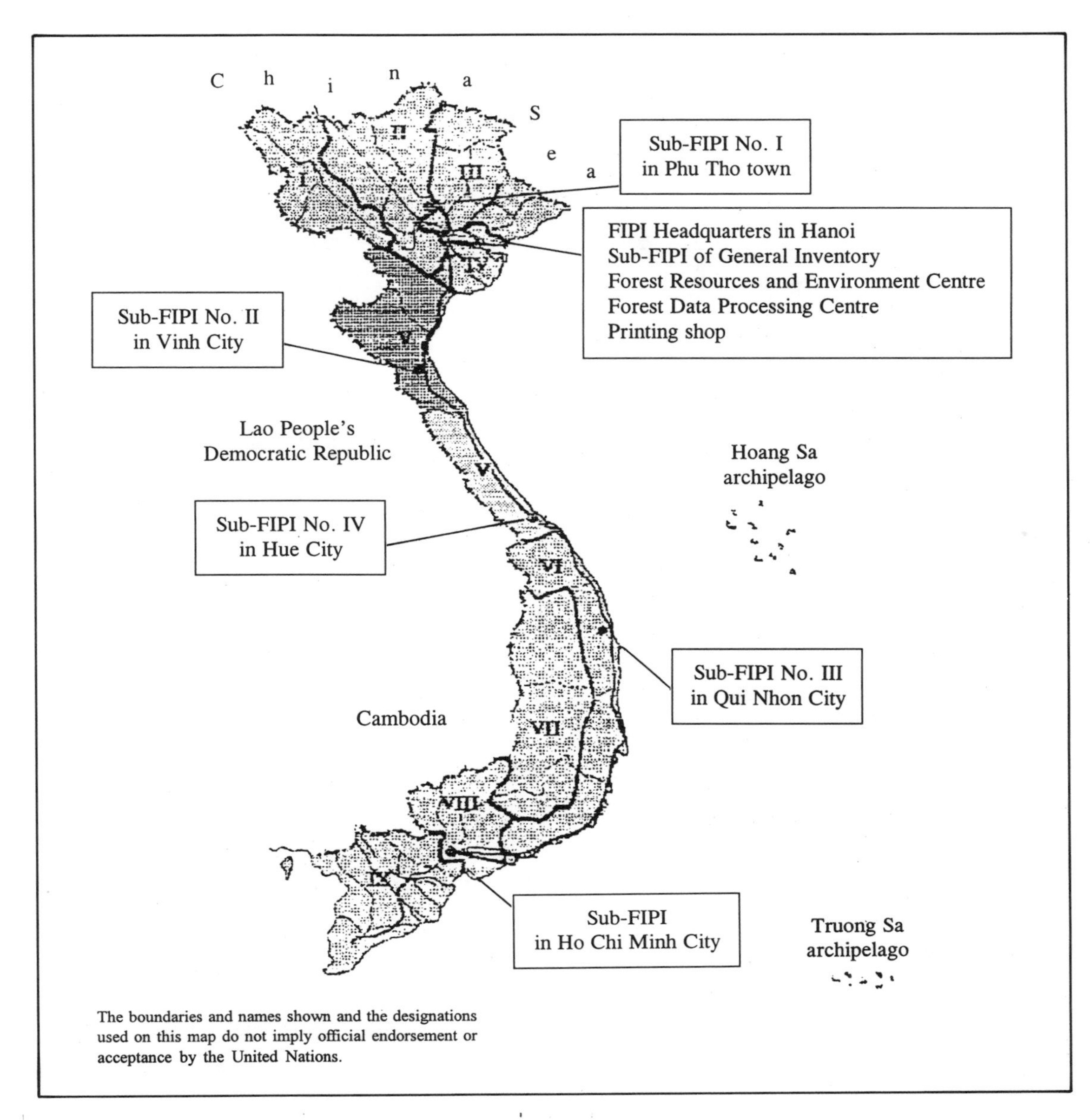

Figure VI. Nine regions of Viet Nam with FIPI centres

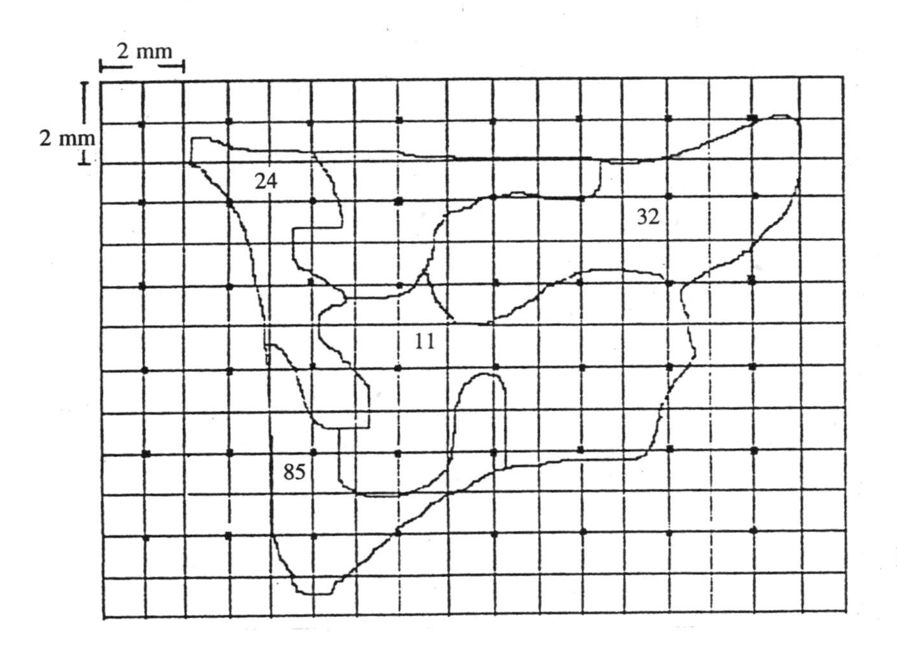

Figure VII. Area calculation by the grid system

VIII. REFORESTATION FOLLOWING AN ECOLOGICAL APPROACH

*Sirin Kawla-ierd**

ABSTRACT

Forest restoration based on phytosociological study, which differs from conventional reforestation, has been attempted in Thailand since 1991. Phytosociological surveys, were carried out to understand the forest ecology of natural and secondary forests. Forest restoration based on the surveys was then applied to restore and manage degraded areas. The six-year results show that the growth performance (basal diameter and total height) of the secondary, late succeessional and climax species indicates their high potential in the restoration of degraded areas.

Introduction

The human impacts on land use and forest exploitation in Thailand and its Asian neighbours are serious problems. The misuse of forest lands and deforestation in the past and especially nowadays causes erosion, floods, drought, loss of biodiversity and soil fertility, climatic changes and so on. Strategies for rehabilitation of tropical forest ecosystems are little known but are most urgently needed.

By definition, "restoration" refers to the recreation, reconstruction, recovery or return of an ecosystem to its original pre-damaged condition, with dominance by a group of native organisms that are within the natural limits for the structure and function of the ecosystem for the local geographic area (Cairns and Buikema 1984; Howell 1986).

For the tropics, restoration with fast-growing species (*Eucalyptus* spp., *Acacia* spp.) is the most common choice. Moreover, monoculture is more popular than multiculture, but it leads to ecological imbalance. A good illustration is even-aged stands. Besides, such plantation forests are easily damaged by insects and diseases.

In 1991, Kazue Fujiwara, Akira Miyawaki and Shunji Murai, in cooperation with the Office of HRH Princess Maha Chakri Sirindhorn's Projects, Chitralada Palace, Bangkok, launched the Re-Green Movement (RGM) in Thailand. The principle of RGM is to foster the creation of native forests that have high biodiversity. This perspective is subject to multifunctions with three major factors. These are, first, the biological objective of conserving species and a green environment for humanity; second, the physical objective of conserving the water balance, to protect against soil erosion and to prevent fire; and third, the chemical aim of producing oxygen and reducing air pollution.

A. Objectives

The objectives of the study are the following:

(a) To develop technology for restoration of tropical forest ecosystems;

(b) To recreate native forests based on ecological approaches (understanding natural forest ecosystems);

(c) To upgrade deforested areas;

* Office of HRH Maha Chakri Sirindhorn's Projects, Chitralada Palace, Bangkok, Thailand.

(d) To educate and promote the participation of school students and local people in interpreting forest values and the impacts from deforestation.

B. Study area and methods

1. Study area

(a) Ban Bor Wee Suan Phung District, Ratchaburi Province

Ban Bor Wee is located in the west of the country (180 km west of Bangkok), about 10 km from the Thailand-Myanmar border. It was established in 1958 by Karen immigrants from Myanmar. It is surrounded with hills. The total area of the village is 16.8 sq km (1,680 ha). The village is bordered by Ban Tha Phak (to the north), Ban Tha Thian Thong (to the south-east) and Myanmar (to the west). The annual precipitation of this village is about 1,200 mm. Average temperature is approximately 28° Celsius. In addition, the elevation is 200 m above mean sea level.

2. Methods

(a) Phytosociological survey

The phytosociological field survey has advantages not only for recognizing and defining plant communities, but also for comprehensive recording of vegetation samples as basic units of natural environments and various scientific studies. The popular method for field survey, classification and description of vegetation is based on Braun-Blanquet and the "Tuxen school" of phytosociology. The key points deal with (a) the selection of optimal field survey locations, (b) measurement of the "total estimate" (cover plus abundance), sociability, and their relationship, (c) the correct method of tablework and its simplification, (d) description of plant communities and (e) development and application of the results (Fujiwara 1987). From analysis and classification of field surveys, scientists move on to species selection for rehabilitation of ecosystems.

(b) Experimental restoration

Seeds of canopy trees were picked up during harvesting time and were sown into a seed bed that was filled with a fertile medium. After germination (about one month), baby seedlings were transferred to plastic pots 11 cm in diameter, which were filled with fertile soil. All pots were placed in a nursery. After six months, seedlings were planted on sites prepared by plough and compost fertilizers. Dense (two or three seedlings per sq m) and random planting were applied. An important step is that seedings were soaked with water before planting and mulched with rice straw and grass after planting.

C. Results: Growth performance of plantation forest

The height of planted seedings increased by almost 1 metre per year and the basal diameter by 1 centimetre per year. The survival rate of the plantation forest is about 85 per cent. The viability of planted seedlings is also good and problems from insects and disease have hardly occurred.

D. Discussion

Growth of plantation stands provides some indication of possible performance in ecological restoration. Furthermore, the speed of restoration with native species into a climax plant community is much faster than in natural succession.

Local participation is essential and is the key to the success of restoration. The management and maintenance (for example, weed control and fire prevention) have the participation of rural clients. Through educational processes, local people become willing to participate in restoration activities. The target group of participants, not limited in age and sex, includes school students, parents, villagers and people outside the village. As a result, the cost of restoration is low.

Weeds spread very rapidly and occupied large areas in the rural rehabilitated zone. On the other hand, few weeds grew in the urban restoration area.

The conservation of soil strongly affects ecological restoration (both natural and artificial restoration). It takes a very long time (more than 5-10 years) for the soil layer to be restored. For this reason, all activities that degrade topsoil must be strenuously avoided.

Mound construction is needed for rehabilitation in Bangkok areas. Since the topography of Bangkok is flat and low (about 2 m above mean sea level), the water table is high. Besides, Bangkok has an inadequate drainage system.

Restoration of degraded land requires an integration of knowledge and skill. It is necessary to have not only an understanding of the forest ecosystem, but also social and economic knowledge.

Restoration should be implemented both for natural forests and buffer forests. The reason is that local people still need the forest products for fuel and other domestic purposes. The restoration of the evergreen forest is important for the ecosystem. On the other hand, the secondary forest is necessary for the use of local people.

The production of seedlings is limited because of the lack of fruit/seeds of canopy species. For example, the fruiting of Dipterocarpaceae species is irregular and the time of seed viability is very short (10 days). Therefore, seed bank/gene pool is one of the effective means of eliminating the problem of seed shortage.

Environmental education is essential to protect, conserve and recreate ecological landscapes and to ensure extensive land utilization for sustainable development.

Bibliography

Cairns, J., Jr. and A.L. Buikema, Jr., 1984. *Restoration of Habitats Impacted by Oil Spills* (Boston, Butterworth Publishers).

Fujiwara, K., 1987. Aims and methods of phytosociology or "vegetation science". In *Papers on Plant Ecology and Taxonomy to the Memory of Dr. Satoshi Nakanishi,* pp. 607-628.

Fujiwara, K., 1993. Rehabilitation of tropical forests from countryside to urban areas. In *Restoration of Tropical Forest Ecosystems* (Kluwer Academic Publishers), pp. 119-131.

Howell, E.A., 1986. Woodland restoration: an overview, *Restoration and Management Notes,* 4(1):13-17.

Miyawaki, A., K. Fujiwara and M. Ozawa, 1993. Native forest by native trees: restoration of indigenous forest ecosystem (reconstruction of environmental protection forest by Professor Miyawaki's method), *Bulletin of Institute of Environmental Science and Technology* (Yokohama National University), 19(1):73-107.

Miyawaki, A. and R. Tuxen, 1977. Vegetation science and environmental protection. In *Proceedings of the International Symposium in Tokyo on Protection of the Environment and Excursion on Vegetation Science through Japan* (Maruzen Co., Ltd.).

Murai, S., ed., 1991. *Applications of Remote Sensing in Asia and Oceania: Environmental Change Monitoring* (Hong Kong, China, Asian Association on Remote Sensing).

Sabhasri, S., 1984. Human impact on the vegetation of Thailand. In A. Miyawaki and others, eds., *Vegetation Ecology and Creation of New Environments.*

Proceedings of the International Symposium in Tokyo and Phytogeographical Excursion through Central Honshu (Tokai University Press).

Santisuk, T., 1988. *An Account of the Vegetation of Northern Thailand. Geoecological Research,* vol. 5 (Franz Steiner Verlag Wiesbaden Gmbh, Stuttgart Germany).

Santisuk, T., and others, 1991. *Plants for Our Future: Botanical Research and Conservation Needs in Thailand* (Royal Forest Department, Bangkok).

IX. DEGRADATION OF THE UPPER MAHAWELI CATCHMENT AND ITS IMPACT

*H. Manthrithilake**

A. History

The whole hill country of Sri Lanka was dramatically opened up in the nineteenth century by the British. The land that was not under human influence was called "wasteland" and declared the property of the British Government. Then it was sold to expatriates (the majority of whom were from Britain) for plantation agriculture. Land in large chunks was sold for a very low price. The land was covered with indigenous, natural species and was a sanctuary for many kinds of wildlife, including elephants.

People who bought the land cleared the jungles very haphazardly, devastating not only the natural wild habitat, but also other resources like land and water. Roads were constructed, settlements were established and crops were grown: first coffee, then tea.

Slope, or elevation, was not a barrier for any of these activities. Trees were burnt; animals were killed; slopes were cleared, bringing down soil; rivers were full of mud; the Kelani River was originally used to transport goods and produce, but because of siltation this became impossible. So planters requested the colonial government in Ceylon to construct a new railway along the Kelani River, which was built with a narrow gauge. The Director of Kew Gardens, London, who visited Sri Lanka in the early 1890s, protested to the British Government after seeing the destruction going on in the hills of Ceylon. As a result, the Secretary for Colonial Affairs appointed a commission to look into the matter and make recommendations. One recommendation was to prohibit clear felling above 5,000 feet, which was adopted promptly. Subsequently, there were several commissions and sessional papers that raised the issue, finally resulting in the Soil Conservation Act of 1951.

Nevertheless, the damage continues to occur and in certain places more than 100 cm of top-soil has been washed off. The consultant of Nedeco, in 1984, estimated that about 1,500,000 tons of soil were transported past the Peradeniya bridge during the period 1952-1982. An estimated average of 35 cm of soil layer had been washed out of the Upper Mahaweli Catchment (UMC) area during the last 100 years.

In 1956, for the first time in Sri Lanka, land-use maps were prepared. This land-use mapping was more biased towards resources available, particularly timber, as can easily be seen from the legend.

Political changes in 1956 and subsequent policy changes have seriously influenced the land-use tenure problems in Sri Lanka. This kind of political influence had another change of direction in 1977 and then from 1990 onwards.

In the early 1990s the Mahaweli Authority carried out land-use mapping in the UMC area for catchment conservation purposes. These maps are at 1:10,000 scale and the legend is biased towards the soil cover by trees, for obvious reasons. Table 1 shows the 1956 and 1991 land cover brought to the legend and uses the same scale, 1:50,000, with differences shown in amount and percentage.

Careful inspection of this table will show that despite the fact that tea has come down in extent, tree cover has increased. This has been verified by the AGA Division.

* Mahaweli Authority of Sri Lanka.

Table 1. Land use in 1956 and 1991

	1956		1991		Difference	Percentage
Tea	116 957.59	(37.6%)	68 444.2	(22.0%)	–48 513.4	–41.5
			(28.5% VP)	–		
Garden	39 738.43	(12.8%)	59 208.5	(19.0%)	+19 470.1	+49.0
MEC	–	–	–	–	–	–
Coconuts	2 555.99	(0.8%)	1 321.6	(0.4%)	–1 234.4	–48.3
Rubber	3 066.74	(1.0%)	10.5	(>0.1%)	–3 056.3	–99.7
Mulberry	0	–	59.9	(>0.1%)	+59.9	–
Cocoa	2 299.71	(0.7%)	3.6	(>0.1%)	–2 296.1	–99.8
Paddy	24 928.88	(8.0%)	20 984.5	(6.7%)	–3 944.4	–15.8
Chena/veg.	36 555.85	(11.8%)	27 448.2	(8.8%)	–9 107.6	–24.9
Grassland	29 736.47	(9.6%)	21 459.0	(6.9%)	– 8 277.5	–27.8
Scrub	241.08	(0.1%)	16 370.5	(5.3%)	+16 129.4	–
OWL+Bam	21 161.70	(6.8%)	17 193.4	(5.5%)	–3 968.3	–18.7
DWL	23 858.84	(7.7%)	38 303.5	(12.3%)	+14 444.7	+60.5
Forest plantation	6 225.03	(2.0%)	21 003.3	(6.8%)	+14 778.3	+237.4
NAL	101.22	(>0.1%)	936.6	(0.3%)	–	–
Rock outcrop	–	–	1 792.8	(0.6%)	–	–
Set/URB	1 829.54	(0.6%)	1 671.6	(0.5%)	–	–
Set	–	–	6 612.5	(2.1%)	–	–
Flooded/water	1 774.28	(0.6%)	8 288.2	(2.7%)	+6 513.9	+367.1
Totals:	311 031.35	(100.1%)	311 112.4	(99.9%)	–	–

Sources: For 1956, *Hunting Surveys*; for 1991, EFCD/FORLUMP maps.

B. Land use

Recent population statistics from Division Resource Profiles (circa 1991) indicate population density in UMC to be 555 people per sq km. Given this high population density, coupled with a shift towards a market economy (and subsequent increased pressures on natural resources), it is perhaps logical to expect increasing land degradation and deforestation. This perception is strongly backed up by anecdotal information from UMC inhabitants. However, if we examine two land-use surveys carried out in UMC 35 years apart, a different picture emerges.

Land degradation occurs in response to many causes, including natural, physical and anthropological activities. Measuring the level of degradation is a very complex task. The word "degradation" itself implies a quality drop from one level to another. The scale of degradation, too, is very subjective. Different experts could assess the same land at various levels of degradation. Land considered most unsuitable on the one expert could be considered very suitable by another. This difference in opinion is due to the fact that land can be put to as many uses as one can think of.

From the agricultural perspective, effects of land degradation could be directly linked to the natural fertility levels of the land. In the absence of fertility information, one could somehow link them to yield levels. Yield characteristics in its wider meaning (crop, water, sediment and so on) could have seasonal as well as long-term trends. Again, absence of such data makes the task impossible.

Land cover could be treated as another indicator of degradation assuming that climate and anthropological practices remain unchanged. Land cover could be taken as a cumulative result of fertility levels, natural quality of soils and climate, and most importantly, anthropological impact. In this context, land-cover assessment too could be subjective, depending on the level of expertise used to interpret the aerial photos or field checking. But if a very limited group of people is involved, this assessment could be fairly accurate.

In the given exercise the following land-use categories were grouped together:

(a) Degraded lands:

- Tea land with less than 60 per cent cover on all slopes
- Grasslands[1]
- Forest plantations[2]
- Rock outcrops

(b) Degrading lands:

- All land uses on slopes of more that 60 per cent except natural forest
- Annual crops except paddy lands
- Urban/settlement on slopes of less that 30 per cent
- Scrub lands

(c) Good lands:

- Good tea cover
- Natural forest
- Perennial crops with good density (KHG)
- Annual crops
- Scrub lands on slopes of less than 30 per cent
- Unproductive/urban

The above aggregation is subjective but lies within LUPPD-specified guidelines. However, in the absence of any direct or objective information on soil/land quality this seems to be the most accurate first attempt (see also the figure and table 2).

Table 2. Results of land degradation monitoring using GIS

Degraded lands			Degrading lands			Good lands		
Classes	**Extent (ha)**	**Per cent**	**Classes**	**Extent (ha)**	**Per cent**	**Classes**	**Extent (ha)**	**Per cent**
<60 per cent; tea cover	25 906	8.33	Slope >60 per cent; land cover except natural forest	6 278.20	2.02	Natural forest	55 567.54	17.86
Grasslands	21 304.09	6.85	Slope >30-60 per cent; annual crops except paddy	11 729.91	3.77	Good tea cover	52 776.07	16.97
Forest plantations	21 302.15	6.85	Slope >30-60 per cent; scrub lands	1 672.60	0.54	Annual crops	34 261.47	11.01
Rocks	1 808.35	0.58	Slope >30-60 per cent; urban and settlement areas	7 345.00	2.36	Other perennials	50 006.55	16.07
						Scrub lands	5 637.15	1.81
						Unproductive lands	7 192.51	2.31
Total	70 320.59	22.61		27 025.71	8.69		205 441.29	66.03

This aggregation was made possible because of 1:10,000-scale land use and slope information available on a computer-based GIS.

[1] Normally, dry grasslands on hill slopes are considered degraded lands in Sri Lanka.

[2] Such grasslands and marginal tea lands were planted with forest with a similar assumption as in note 1.

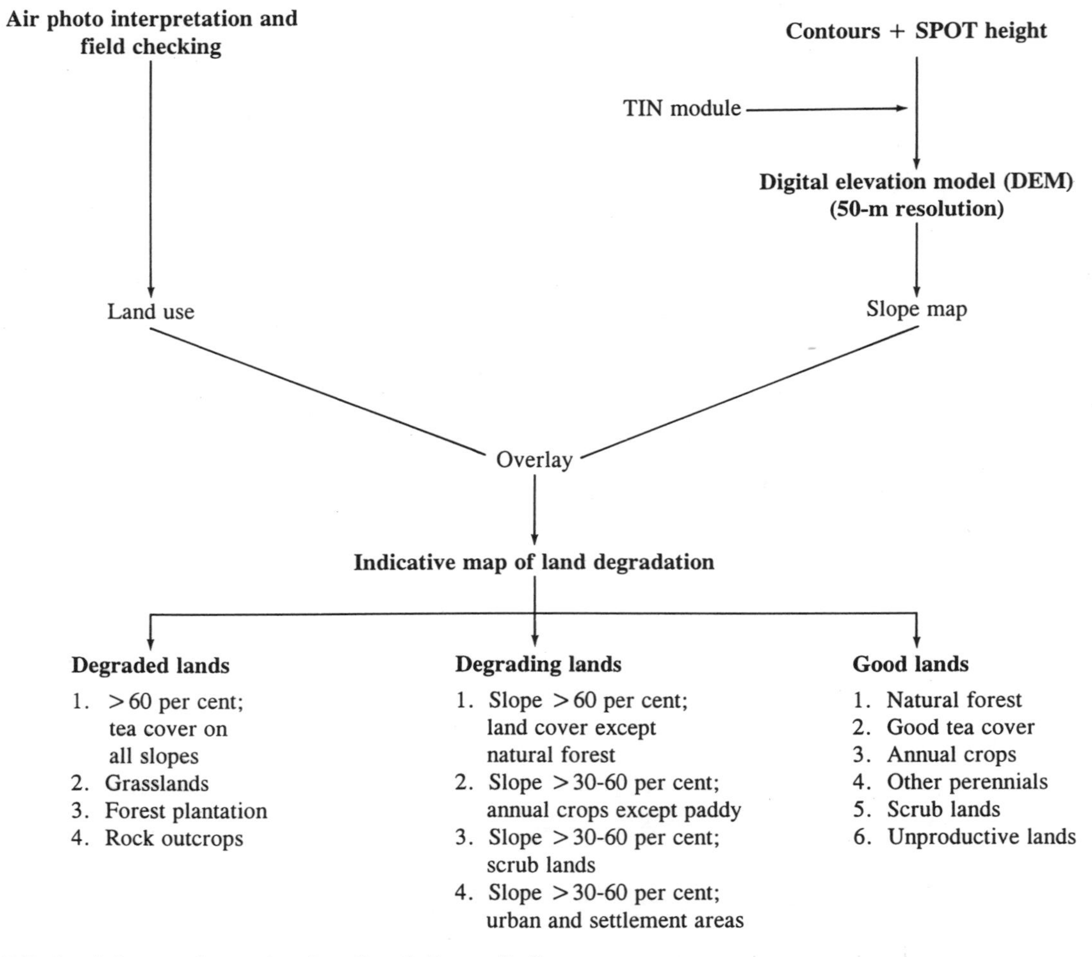

Methodology of monitoring land degradation

X. MANAGEMENT OF ECOSYSTEMS IN MYANMAR

Ohn Gyaw

A. Introduction

Myanmar has long coastlines, a considerable amount of low-lying areas, hilly regions, forested coverage of the country over almost three quarters of the total surface area, paddy growing agricultural practice and other features. The dry zone of the country is in the central part of Myanmar, where annual rainfall is about 800 mm. The isohyetal map for annual normal rainfall of the country is shown in figure I. Although the country is generally blessed with abundant water, this resource is poorly distributed in both space and time. The heavy rains during the south-west monsoon and the torrential downpours associated with sudden storms lead to sustained flooding in the wetter areas and to flash floods in the dry parts or in places where steep mountain torrents overflow.

The country is divided topographically into four regions. The eastern Shan Plateau is a highland region averaging 900 metres in height and merging with the Dawna Range and the Tanintharyi Yoma towards the Isthmus of Kra. The central belt spans the valleys of the Ayeyarwady, Chindwin and Sitthang rivers with a mountainous region in the north and a vast, low-lying delta in the south that covers an area of 25,900 sq km. It produces almost all the nation's rice. The western mountain belt, also known as the Rakhine Mountains, is a series of ridges that originate in the northern mountain area and extend southward to the south-western corner. The Rakhine coastal strip is a narrow, predominantly alluvial belt lying between the Rakhine Mountains and the Bay of Bengal. In some places the strip disappears as the mountain spurs reach the sea. In the offshore area there are hundreds of islands, many of which are cultivated.

The most striking feature in the meteorology of Myanmar is the alternation of seasons known as the monsoon. Strictly speaking, monsoons are seasonal winds whose directions reverse twice during the year. Lying within the tropics, with the great Asian continent to the north and the wide expanse of the Indian Ocean to the south, Myanmar furnishes one of best examples of a monsoon country. During the winter months of the year from December to February the general flow is from the north or north-west in the northern parts and from the north-east in the rest of the country. In this season the air over the country is mainly of continental origin and hence of low humidity and low temperature, and the season is known as the north-east or winter monsoon season. In the summer months of May to October the general flow of wind is from the opposite direction, from sea to land, with the tropical maritime origin, so the season is one of much humidity, cloudiness and rain. The direction of winds in the Bay of Bengal and the Andaman Sea being south-westerly, the season is named the south-west or summer monsoon season. Between these two principal seasons are the transition seasons of the hot and dry weather months, March, April and May, and the retreating monsoon months of October and November. It is noteworthy that most storms which reach Myanmar's coasts occur mainly in the transition seasons.

B. Greening project for the nine critical districts of the arid zone of central Myanmar

1. Description of the arid zone areas

The nine critical districts of the arid zone of central Myanmar suffer intense heat of up to 43°C in the summer with annual precipitation of about 500 mm to 1 m which is unevenly distributed throughout the year. The arid zone has geographically and geologically supported sparse forest growth and it has been cut intensively for use as fuelwood. The arid zone has one third of both the population and area of Myanmar, where 11.3 million people, or 81 per cent of the population, are rural. It is one of the most important agricultural areas in the country, producing the major cash crops, as well as supporting half the national cattle population. Agricultural land productivity is negatively affected by population pressure, cropping on inherently poor and fragile soils, low input use and environmental deterioration due to deforestation for fuelwood and wood supplies. The area of the greening project is shown in figure II.

2. Environmental concerns

A major human concern is the acute shortage of fuelwood, a basic and essential commodity, compounded by environmental deterioration in critical areas in the arid zone. The dry zone is in crisis, as there has been a sharp decline in fuelwood supply, underground water supply, farm environment and stagnating agricultural production over the last decade. Dry season water resources are declining because of decrease in groundwater recharge and increased surface run-off from the degraded lands. Uncontrolled water run-off leads to low moisture retention in the subsoil and soil erosion. Devoid of tree cover and shelter belts, croplands are exposed to desiccating southerly winds during the summer, resulting in the loss of topsoil. The acute imbalance between forage/fodder supply on the one hand and livestock population on the other also exerts strong environmental pressures.

3. Management undertaken by the government

Starting in the 1980s, the planting tempo increased to around 40,000 acres per year, and it increased by about 10,000 acres each year; now the planting programme target is fixed at around 80,000 acres annually. A total of 78,000 and 77,000 acres of plantations were established in the years 1992-1993 and 1993-1994, respectively. It was stipulated that out of the total plantation areas, 40 per cent would have to be fuelwood plantation in certain areas to meet the fuelwood needs of the indigenous people.

Apart from the regular plantation establishment of about 80,000 acres annually, the Forest Department has been distributing around 4.5 million seedlings free of charge to local communities and governmental organizations in an attempt to raise wood lots and for planting under the agroforestry system in farmers' own holdings and farmyards. This free seedling distribution scheme was started in 1977/78 to encourage greening in non-forest areas such as homesteads, farmlands and along shelter belts. The practice gained support and in 1992/93 and 1993/94 fiscal years about 11 million were distributed annually. Starting from 1993/94, this Nationwide Afforestation Campaign of seedling distribution and planting is being undertaken not only in July, but throughout the monsoon season. In 1994/95, over 10 million seedlings were handed out in the planting period with the purpose of upgrading the greening of the country.

In 1991 a National Commission for Environmental Affairs was established as a focal point to formulate a national environmental policy framework and to coordinate the various line ministries in the execution of policy guidelines, directives and functions concerned with the environment. The Government of Myanmar has signed an accord to participate in the Global Environment Facility and is also signatory to the Convention on Biological Diversity and the United Nations Framework Convention on Climate Change.

Recognizing the crucial role of grass-roots involvement and the overriding need to promote synergetic linkage between public and private initiatives, the government enacted a new Forest Law in 1992, which encourages community participatory fuelwood and commercial planting activities.

This project addresses the greening effect, the acute fuelwood shortage and environmental degradation in 42 townships of the nine critical districts of the arid zone of central Myanmar. Environmental degradation has a major impact on groundwater supply, the health and nutrition of the rural poor, and their income earning opportunities. The project output will reduce forest depletion and degradation and enhance the prospects for environmental conservation, as well as meet the fuelwood demand of the local people and improve the livelihood of the rural poor.

C. Conclusion

The government is doing its utmost in collaboration with United Nations agencies in order to implement the Dry Zone Greening Project, Ayeyarwady Mangrove Forest Village Development Project and Multi-purpose Village Firewood Plantation Project. These projects are mainly concerned with environmental conservation and at the same time attempt to undertake proper management of ecosystems in the country.

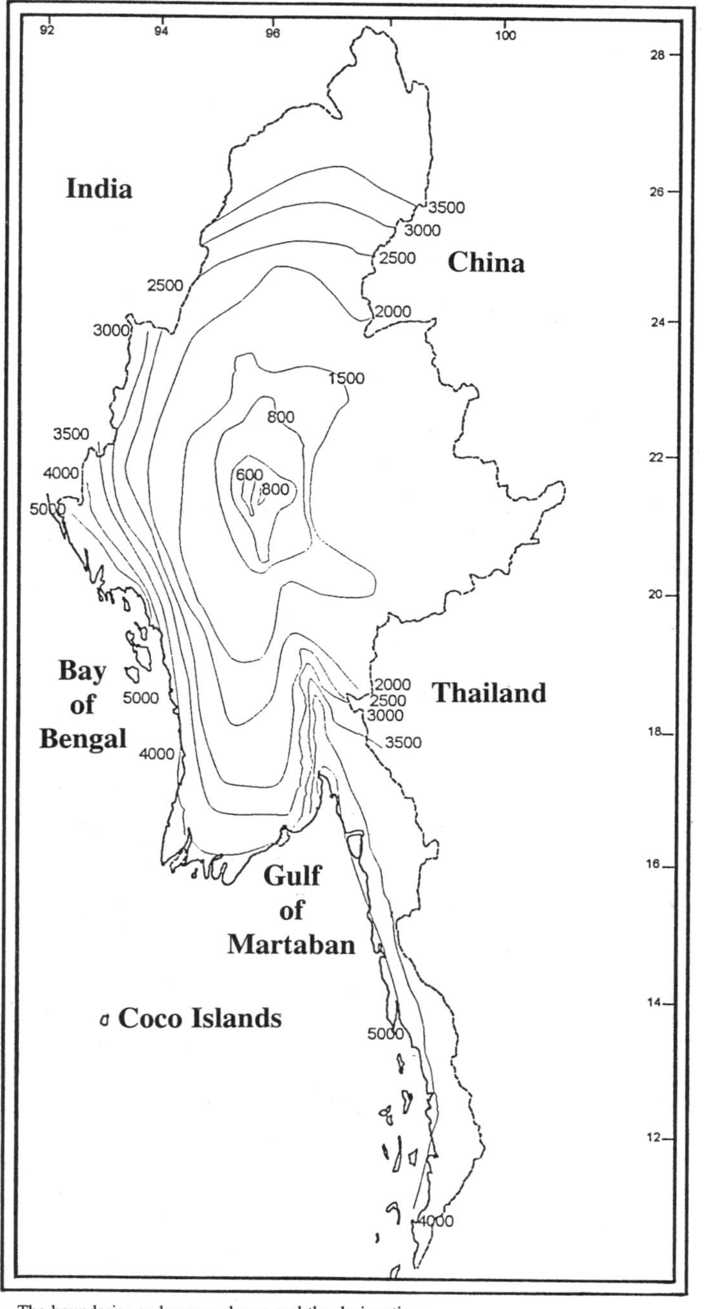

The boundaries and names shown and the designations used on this map do not imply official endorsement or acceptance by the United Nations.

Figure I. Isohyetal map for annual normal rainfall of Myanmar (mm)

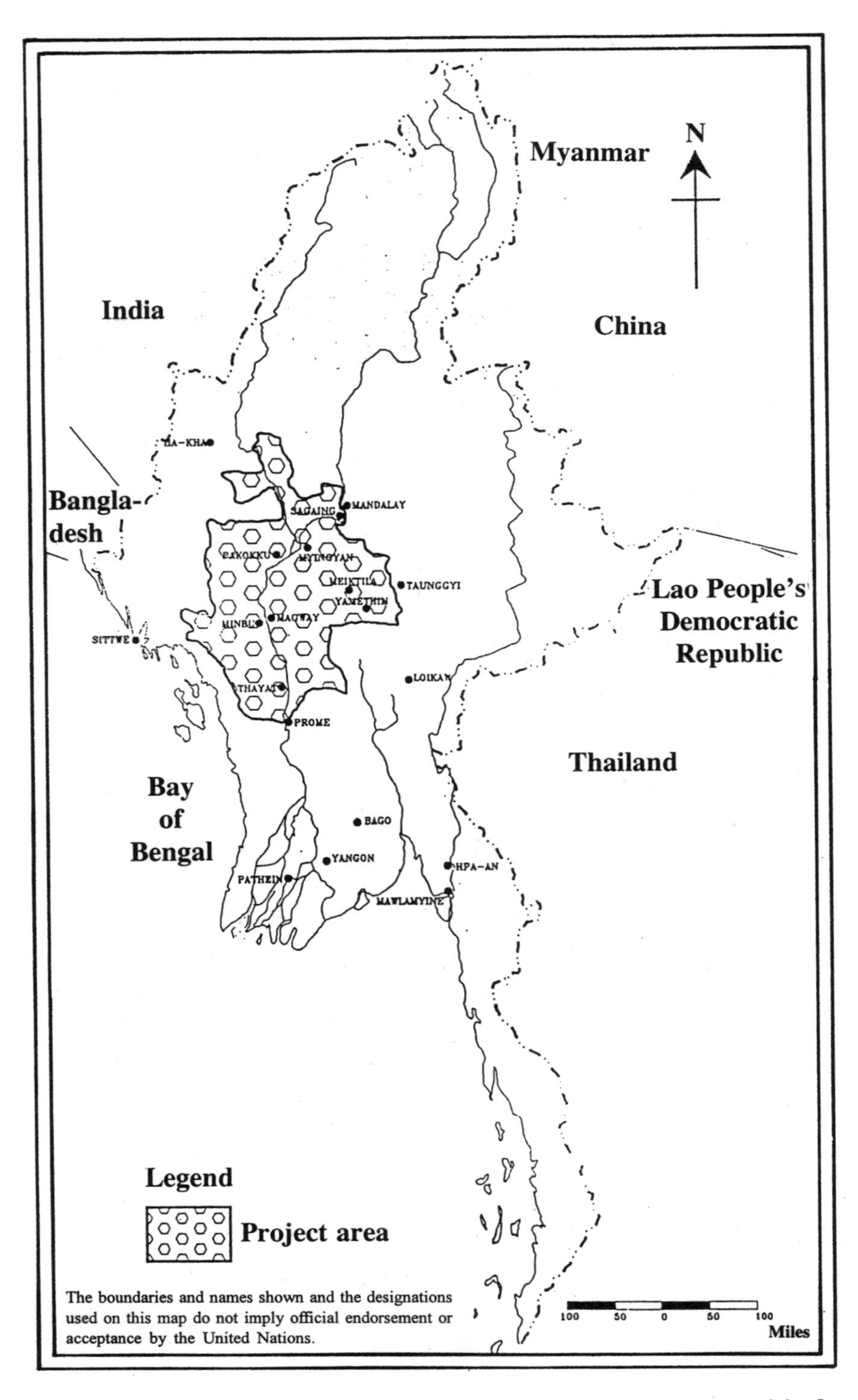

Figure II. Area of the greening project for the nine critical districts of the arid zone of central Myanmar

PART FOUR

PAPERS: MICROWAVE REMOTE SENSING

XI. EUROPEAN REMOTE SENSING SATELLITE, SIX YEARS IN ORBIT: THE MISSION AND SELECTED APPLICATIONS

*Robert Schumann**

A. The ERS instruments

The synthetic aperture radar (SAR) image mode supplies high-resolution radar imagery over land, sea and ice. High-quality information acquired from many regions of the world has been exploited in numerous areas using techniques at the forefront of digital image processing.

The SAR wave mode acquires SAR "imagettes" of the ocean every 200-300 km along the satellite track, which are then used to derive the direction and period of ocean waves.

The scatterometer uses three antennas to measure small-scale sea surface roughness and thus resolve surface wind speed and direction over the world's oceans. The data are also exploited to derive physical parameters over sea ice, ice shelves and land.

The radar altimeter measures the distance from the satellite to the Earth's surface with centimetric precision. In addition, it routinely measures wave height and wind speed over the oceans.

The along-track scanning radiometer (ATSR) measures sea surface temperature with high precision thanks to the innovative design of the instrument calibration. It also supplies atmospheric corrections to the altimeter through microwave radiometer measurements.

The Global Ozone Monitoring Experiment (GOME) measures atmospheric absorption in a section of the electromagnetic spectrum. Stratospheric trace gas profiles (ozone, nitrogen oxides and others) are derived and assimilated over time to generate global monitoring products.

The PRARE and laser reflector provide information on the position and velocity of the satellite to support precise orbit determination.

B. The ERS missions

ERS was designed to cater for a wide range of scientific and operational requirements, for which three different repeat cycles have been applied during the mission lifetime (see figure):

(a) The 3-day repeat cycles were suitable for the commissioning of the satellite instruments as well as for the demonstration over selected areas of monitoring applications requiring frequent update with new data;

(b) The 35-day repeat cycles combine best the requirements for frequent data acquisitions and the capability for total global coverage with SAR;

(c) The two 168-day repeat cycles were dedicated to obtaining a dense sampling of tracks at all latitudes for the altimeter. The data recovered have yielded a precise geoid over sea and ice.

For nine months in 1995-1996, ERS-1 and ERS-2 were operated on the same track one day apart, the Tandem Mission. Over most of the Earth's land surface, pairs of SAR images were acquired, one image from each satellite. These image pairs are suitable for interferometric applications dependent on the relative positions of the satellites at their respective time of acquisition.

---

* European Space Agency representative to South-East Asia, Asian Institute of Technology, Bangkok.

To further the use of ERS data, ESA has implemented several "Announcement of opportunity" (AO) and "pilot project" programmes. The goal of the AOs is to promote scientific research, while the goal of the pilot projects is to encourage operational use of ERS data.

Several scientific symposia and application workshops have been held to allow the presentation of project results to a wide audience.

C. Uses of ERS data

1. Exploitation of natural resources

On land, the use of optical satellite imagery for mineral and hydrocarbon exploration is well established. The viewing geometry and imaging mechanism of SAR yields additional information for identification and classification of geological structures.

Offshore exploration managers use anomalies in the Earth's gravitational field to pinpoint areas that may prove productive, as perturbations in the local gravity field are due in part to variations in rock density, which can indicate oil-bearing structures. Gravity anomaly information has been derived from ERS Geodetic Mission altimetry to provide a global high resolution data set to oil explorers.

Seepage from sub-sea oil-bearing structures can be detected at the surface using an archive of ERS SAR imagery, allowing persistent seepage from the seabed to be identified. This information can be combined with satellite-derived gravity information to narrow further the search for oil.

2. Monitoring the coastal zone

Shallow coastal seas can often be subject to silting, a process that needs regular monitoring, particularly in critical areas such as the approaches to harbours. This monitoring process can be rendered more cost-efficient by the use of bathymetric information derived from the ERS SAR, which decreases the resolution in track spacing required from a ship-based survey.

Oil slicks originating from dumping by vessels or rigs at sea make up 45 per cent of all oil pollution. Aircraft surveillance is required for detection and prosecution of offenders, but the area monitored can be significantly enlarged with simultaneous application of SAR imagery.

3. Hazards and risks in the marine environment

Reliable forecasting and "now-casting" of marine hazards such as waves and sea ice aid the critical decision-making process during operations at sea. Climatological information is applied in project planning to cost bids, to produce environment statistics for design of offshore structures, and to plan operations.

Wind fields from the ERS scatterometer are used to correct atmospheric models for weather forecasting. These wind fields, together with wave height, period and direction from the altimeter and SAR, are input to wave propagation models, generating wave forecasts and now-casts. In addition, the climate of a particular marine area can be derived through archiving of these measurements, allowing both prevailing and extreme characteristics of a region to be determined.

Regional monitoring of sea ice using optical, thermal and passive microwave sensors is well established. SAR imagery provides additional high-resolution information, resolving navigable features such as leads and polynyae, extremely useful in critical areas.

4. Monitoring forestry and agriculture

The rate of tropical deforestation in the world is not known with any precision, but is estimated by the Food and Agriculture Organization of the United Nations (FAO) as around 15.4 million ha per year. SAR data provide a method for deriving maps of forest extent and type in tropical areas that have not previously been monitored by satellite, because of near continuous cloud cover.

SAR data are being used uniquely, or synergistically with other remotely sensed data, to provide inventories of timbered areas, to map the encroachment of farming activity on to forested areas, and to map forest damage.

Security of food supply is of concern in both the developed and developing worlds. Public and private sector organizations have been established to monitor food supply, with a responsibility to forecast local harvests, and to assess the import/export situation in order to control prices and meet the needs of the population. Monitoring of rice crops in South-East Asia is one example of how SAR data can be used to aid the management of agricultural resources.

5. Responding to natural disasters

Globally, damage inflicted by natural disasters kills an estimated one million people each decade and leaves millions more homeless. In this, the United Nations International Decade for Natural Disaster Reduction, the human social and economic costs of disasters are being assessed. The ability of differential SAR interferometry to detect and quantify extremely small surface height variations, such as those occurring before a volcanic eruption, can aid in the management of risks associated with them. Anticipation of and response to flooding events is aided by the use of ERS data, which is particularly efficacious because of the ability of ERS to see through the bad weather conditions that accompany flooding.

6. Map compiling and updating

The increase in development in remote areas of the world is generating a demand for up-to-date maps of topography and land cover. Many areas of the world are poorly or insufficiently mapped, and any existing information is often out of date. Additionally, the rapidly growing market for geographic information systems has increased the demand for digital maps. These trends have led to the widespread use of optical satellite-derived information.

ERS SAR imagery can be used to generate high-quality digital elevation models using SAR interferometry; to provide thematic information on land cover, including tropical areas; to improve localization of products complementing Global Positioning System (GPS) measurements in remote areas; and to rectify old or inaccurate maps or maps generated using other space data to a high localization accuracy.

7. Global change with ERS scatterometer

Ice sheets such as the one that covers Greenland are believed to be a sensitive indicator of global warming. Data from the ERS scatterometer have been used to monitor snow properties, especially snow-melt during the arctic summer. In this way the interannual variability of the snow-melt on the Greenland ice sheet has been monitored since 1992. The melt zones are easily recognizable by a strong decrease in the backscatter values in July. A trend can be observed of increased melting from 1992 to 1995, with a decrease observable in 1996. The 1995 season was clearly an extreme event, resulting in snow melt at high altitudes, where normally the snow would stay dry throughout the summer. This research represents an example of the significant contribution of ERS to the study of the Earth's climate and to the search for evidence of recent changes in climate brought about by man.

8. Surface movement with ERS synthetic aperture radar

Close monitoring of small surface movement is of vital importance, because of the potential hazard it represents and the economic damage it may cause. Examples of such phenomena include earthquakes, volcanoes, landslides, glacier surges, and subsidence due to water or oil withdrawal.

The technique of differential SAR interferometry was used to capture the movements produced by the 1992 earthquake in Landers, California, and by several of its aftershocks. The movement can be seen as fringes where the repetition of a colour through the full colour cycle is the result of terrain movement towards the satellite of half the radar wavelength (i.e. 28 mm).

The technique is somewhat limited in its application in areas with dense vegetation, but represents a unique tool for monitoring surface movement phenomena, with many potential benefits.

9. Detailed ocean floor topography from ERS altimeter

The Geodetic Mission of ERS was made up of two interlaced 168-day repeat cycles to yield a tight grid of altimetric measurements over the world's oceans. This data set has allowed the computation of high-resolution mean sea surface, gravity anomalies and sea floor topography. Within the area of marine geophysics this information is used for lithospheric structure analysis, and studies of plate tectonic kinematics and small-scale upper mantle convection, as well as three-dimensional structure and evolution of mid-ocean ridges. The gravity anomaly product is also of interest for offshore oil exploration to aid the identification of oil-bearing structures. In addition, data from the Geodetic Mission, together with data from similar instruments carried on other platforms, allow the generation of detailed information on the dynamic circulation of the oceans. The improved knowledge of the gravitational field can also be used to facilitate the orbit maintenance of polar orbiting satellites.

10. Oil spill monitoring with ERS synthetic aperture radar

The amount of oil illegally dumped from ships every week is equivalent to a disastrous spill from a supertanker. The scale of the problem dictates the use of effective monitoring and legislative measures in order to ensure the protection of the environment, especially in coastal areas, which are particularly at risk from contamination.

ERS SAR data are well adapted to the detection of surface features such as oil slicks. Images acquired and processed in near real time from the Tromso Satellite Station provide information over a wide area, which can then be used to optimize the flight patterns of aircraft operated by national environment agencies.

Several national pollution control agencies routinely make use of this service, which has proven to be cost effective for increasing the area monitored, complementing the use of aircraft.

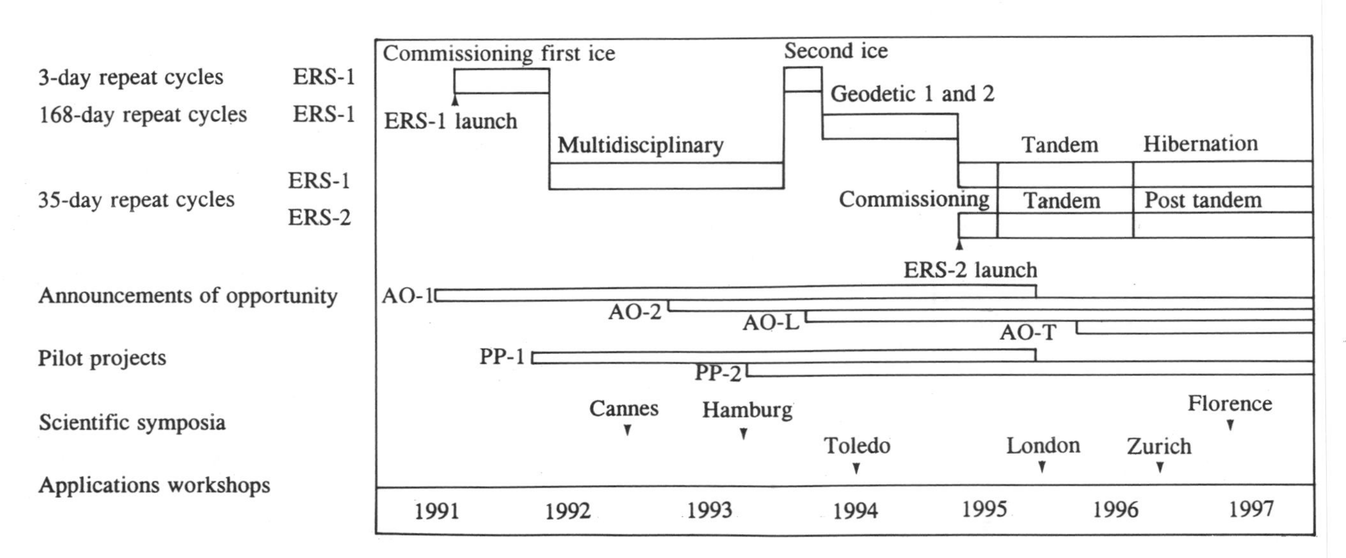

ERS repeat cycles, missions and other activities

XII. SIGNAL PROCESSING OF JERS-1 SAR DATA

*Makoto Ono**

A. Introduction

The advances of computer technology in the last 15 years enable a personal computer to carry out processes formerly requiring a mainframe or supercomputer. Remote-sensing image processing is one of the jobs in this category. This paper introduces a PC-based synthetic aperture radar (SAR) processor to produce SAR images from spaceborne SAR row data. It is quite practical doing this work, because process flexibility – like changing reference function or multi-look numbers for specific image analyses – is important and sometimes essential, while data centres cannot respond to such non-standard requests from various data users. It is also preferable to have the flexibility of changing Doppler frequency weighting or sampling for SAR interferometry. The PC-based SAR processor has the flexibility to answer these requirements.

B. PC program of SAR processor

1. Preface

The computer program on PC is written for Macintosh Power PC with a FORTRAN code. The FORTRAN compiler to be used generates native codes for the Power PC. The compiler is compatible with ANSI 77 standard, so that source code is easy to transport to the other machines if the code does not include the PC-specific human-machine interface process.

We have designed the program to process data in a batch process, in which data input and output are done to input and output data files. In consideration of the limited PC file size or memory size, we have introduced some techniques to reduce the process unit.

2. Process flow

Process flow of the SAR processor is shown in figure I. It is a traditional two-stage FFT-IFFT correlation process for range compression and azimuth compression. The difference in azimuth compression from between the traditional process exists in the range migration correction. The range migration correction is designed to be done on unequally spaced range lines to make equally spaced along-track ground lines. By the design, cross-track mapping on the map does not require non-equal-space resampling. Some other technique is applied to avoid disk access number, so that the PC-based SAR processor can obtain practical process speed. The technique is that the range compressed SAR data of several range lines are merged to make small two-dimensional arrays, each of which is stored as a record. The idea is illustrated in figure II, where R_{1k}, R_{2k},.... R_{nk} represent the small two-dimensional arrays and C_{ij} represents each complex number in the arrays. By doing this, several lines of processed data are read at a time either for range direction or azimuth direction without spending time on corner turns. Azimuth compression programs are in two separate versions to produce either a real image or a complex image, while other parts of the programs are the same for each process.

3. Azimuth reference function

The calculation of the reference function of azimuth compression is another key point to produce high-quality images. We have designed this program to generate the reference function as a time series of slant range to a point target, of which the time origin is zero Doppler time. In this way, the produced image of a range line is mapped as if the synthesized antenna beam centre is

---

* Remote Sensing Technology Center of Japan.

directed to zero Doppler plane, which is perpendicular to the satellite velocity vector on Earth at fixed coordinates. We can also indicate that by this method of reference function generation, there is no Doppler frequency modulation for an interferometry pair of complex number images. The reference function is refreshed every azimuth line.

4. Processed image size and process speed

To run SAR processing programs on a PC, the size of unit process is important in order to achieve a practical speed and obtain practical area size. In the developed program, 4,096 range samples by 8,192 azimuth samples are processed at a time. For the JERS-1 SAR, a 17 km x 20 km area is imaged at a time. This area size is practical both for PC image handling size and local area analysis using this image. Process time and interim file size as well as program code size and required memory size are shown in table 1. Process time is measured on Power Mac 7100, whose clock frequency is 80 MHz. Since there is no specific function of the PC chip that has been used for the SAR processing, similar speed will be achieved on pentium machines.

Table 1. Process time and size of the program and data

Step	Program/ memory/file size	Time (min.:sec.)
Orbit data edit	200 k (program) 60 k (memory)	0:03
Range compression	217 k (program) 2 319 k (memory)	48:50
Azimuth compression	217 k (program) 2 323 k (memory)	68:20
Interim data file	46 097 k (file)	-
Image data file	10 354 k (file)	-

To reduce interim file size, interim data for both correlation processes is scaled to two-byte integers although the processes are performed with floating point number. Output image both real and complex is also scaled to two-byte integer numbers. As for the raw data, CD-ROM is suitable to handle on a PC. More than four 2,400-foot reels of 6,250 BPI CCT data can be copied on a CD-ROM.

5. Some examples of processed image

Using the SAR processing software, I have processed JERS-1 SAR data. Figure III is a real image of the Mt. Fuji area. This is equivalent to the standard three-look image of the NASDA level-2 product.

C. Quick look image generation

1. Programs

For the PC to process a full scene of JERS-1 data is a heavy load even today. Quick look processing with reduced resolution is the best way to relax the load. We have developed a quick look processor to process JERS-1 SAR data in one and a half hours on a Macintosh whose clock frequency is 80 MHz. The technique to reduce data size by sacrificing the resolution is pre-summing. For range pre-summing, a weighted average to perform low pass filtering was adopted. Spectral bands of original SAR raw data were reduced to one fifth of the original band. This reduces signal amount to one fifth of the original, and range resolution is five times larger than the original, which means that about 100-m ground resolution will be achieved. For azimuth pre-summing, phase shifted low pass filtering was adopted so that succeeding four azimuth data makes a phased array antenna, whose beam direction is adjusted to the centre of the actual antenna beam. By this process the raw data size is reduced to one twentieth of the original data. After this step, SAR hardware parameters like spectral band and azimuth antenna size are replaced to meet the pre-sum condition, and the process is the same with the original SAR processor. Example images of the quick look processor are shown in figure IV.

2. Mosaic of SAR images

The SAR image is cloud free. This means that the SAR image can be acquired on a scheduled basis. This is a strong advantage of the SAR system for monitoring purposes. Contiguous generation of a regional mosaic is also feasible. A mosaic of SAR images is not difficult except for mountainous areas. SAR image mapping accuracy depends only on the orbit in contrast to the optical sensor image case, where both orbit and attitude affect the mapping accuracy. Owing to this characteristic, and since the reference function is generated on the orbit information associated with the raw data acquisition, the along-track adjacent scene is just connected to the corresponding point. In our processor, along-track direction is parallel to the sub-satellite locus. Given that the source images are taken on the same path, all the along-track mosaics are performed by overlapping adjacent images without any relative rotation of images. This is true even in the mountainous areas. For cross-track mosaic, relative rotation and foreshortening correction are required because sub-satellite locus is not parallel. As for the correction of foreshortening, the process can be eliminated if the pixel size is larger than 100 m. With the parameter adopted by the JERS-1 SAR design, the relative foreshortening value in near range and far range is less than 100 m even if the height of a mountain is 3,000 m.

D. Interferometry

1. General requirements

Interferometry is one of the most common uses of SAR data. To obtain a well-conditioned interferogram, phase preservation of reconstructed SAR complex data is essential. Phase preservation is realized when the reference function is constructed so that the reconstructed image is aligned to zero Doppler plane and the slant range phase delay is counted on that plane. The developed SAR program satisfies this requirement. When the reference function is constructed on the centre of the antenna beam, and an adaptive process like Doppler centroid introduced, the reconstructed phase suffers the change of antenna attitude. In this case, phase preservation will be hard to maintain.

2. Examples

Using complex output images, we have processed repeat pass SAR interferometry. Pairs of interferometry images are matched only by image pixel position matching. Since both images are mapped on an equal ground space grid, no rotation or size scaling is required. Figure V is an example of fringe for the same area as figure III.

E. Conclusions

PC-based SAR processing programs are developed with practical processing speed and practical area size. This program is useful for some local area analysis with a PC. In addition, the price of raw data of JERS-1 SAR is about one tenth of the price of level-2 image data in Japan. Processing it oneself is a quite economical way to obtain SAR images.

This program will be uploaded on the RESTEC World Wide Web server as freeware. Already, a real data generation version is uploaded. A Windows version will be uploaded soon. As for the interferometry kit, we are now preparing to convert the programs into C language. This version will be uploaded around the summer of 1998.

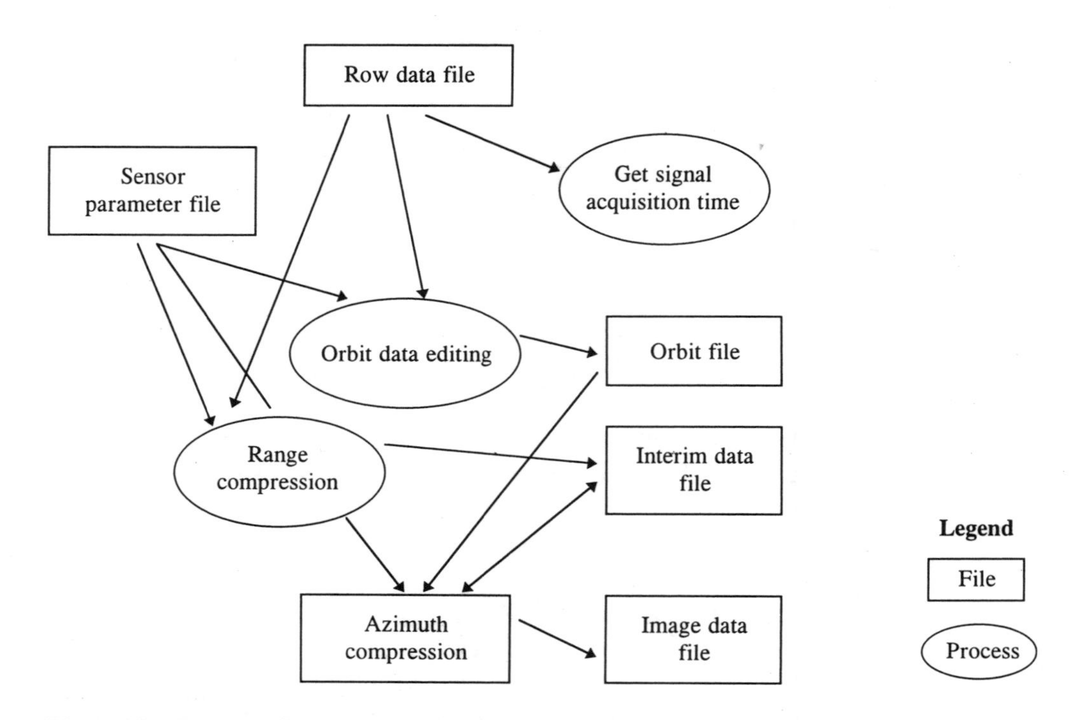

Figure I. Process flow of the PC SAR processor

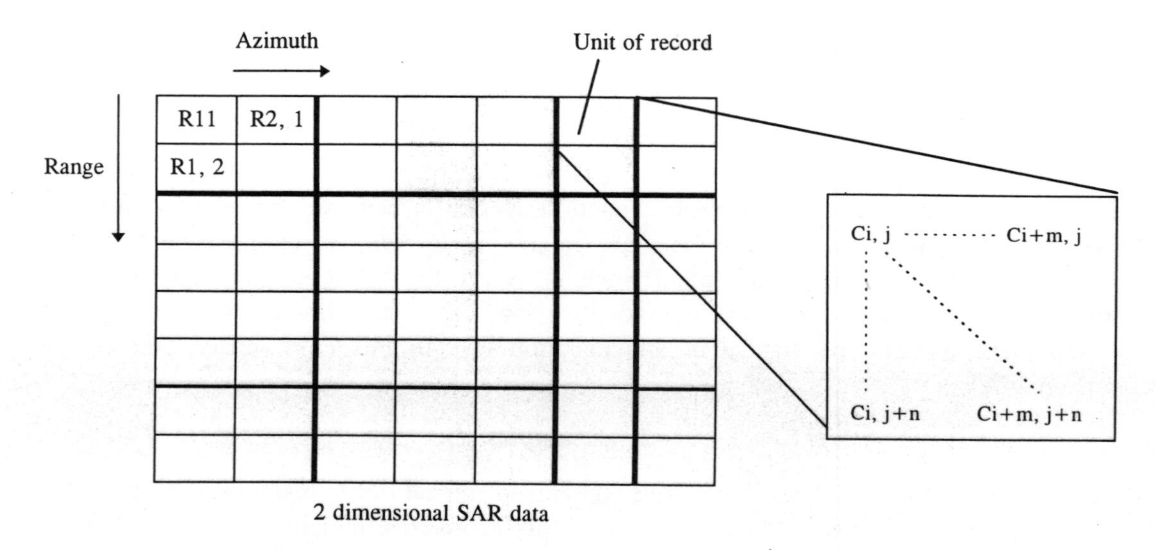

Figure II. Structure of interim data file

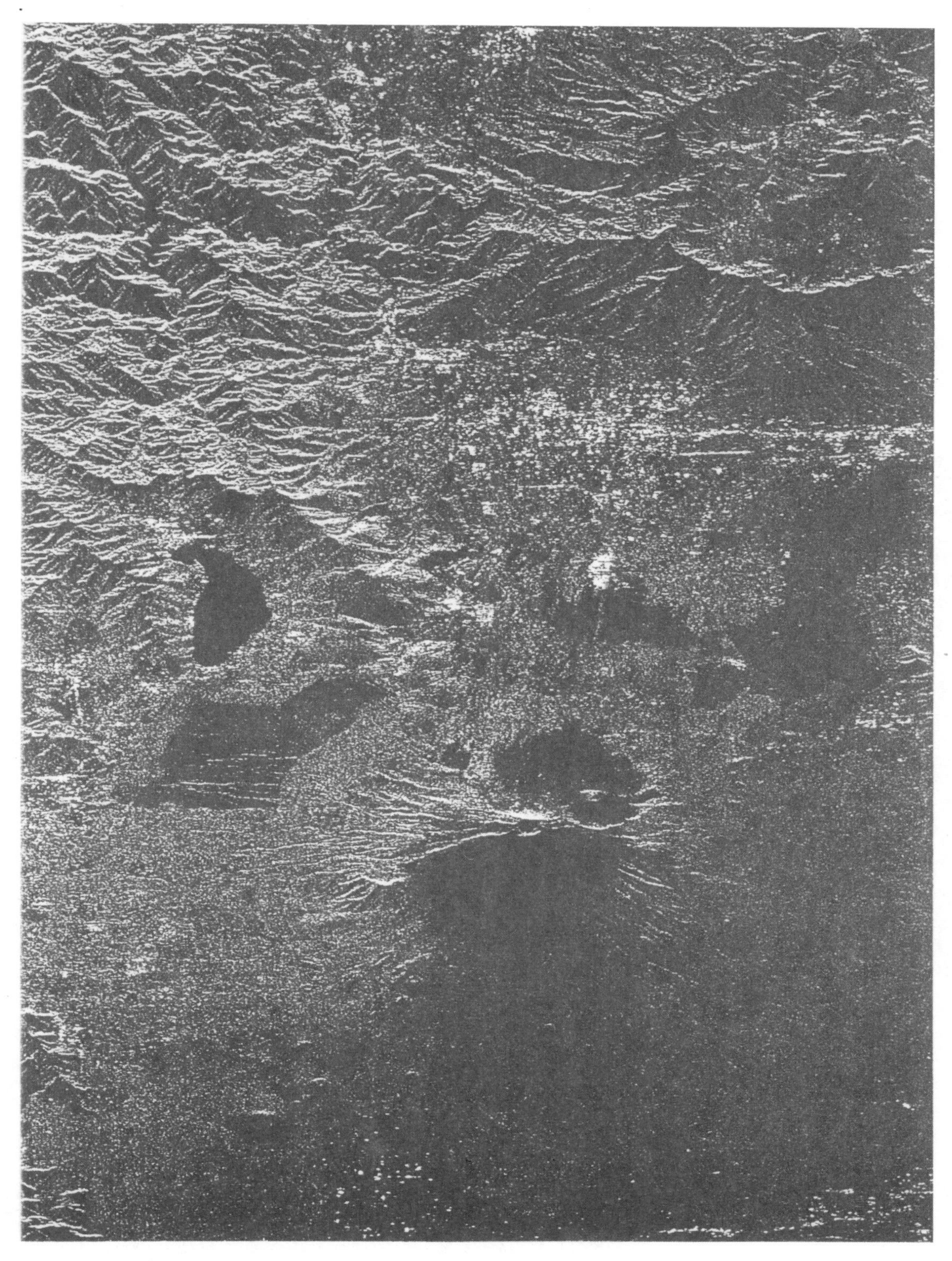

Figure III. Processed image (Mt. Fuji 7 July 1993)

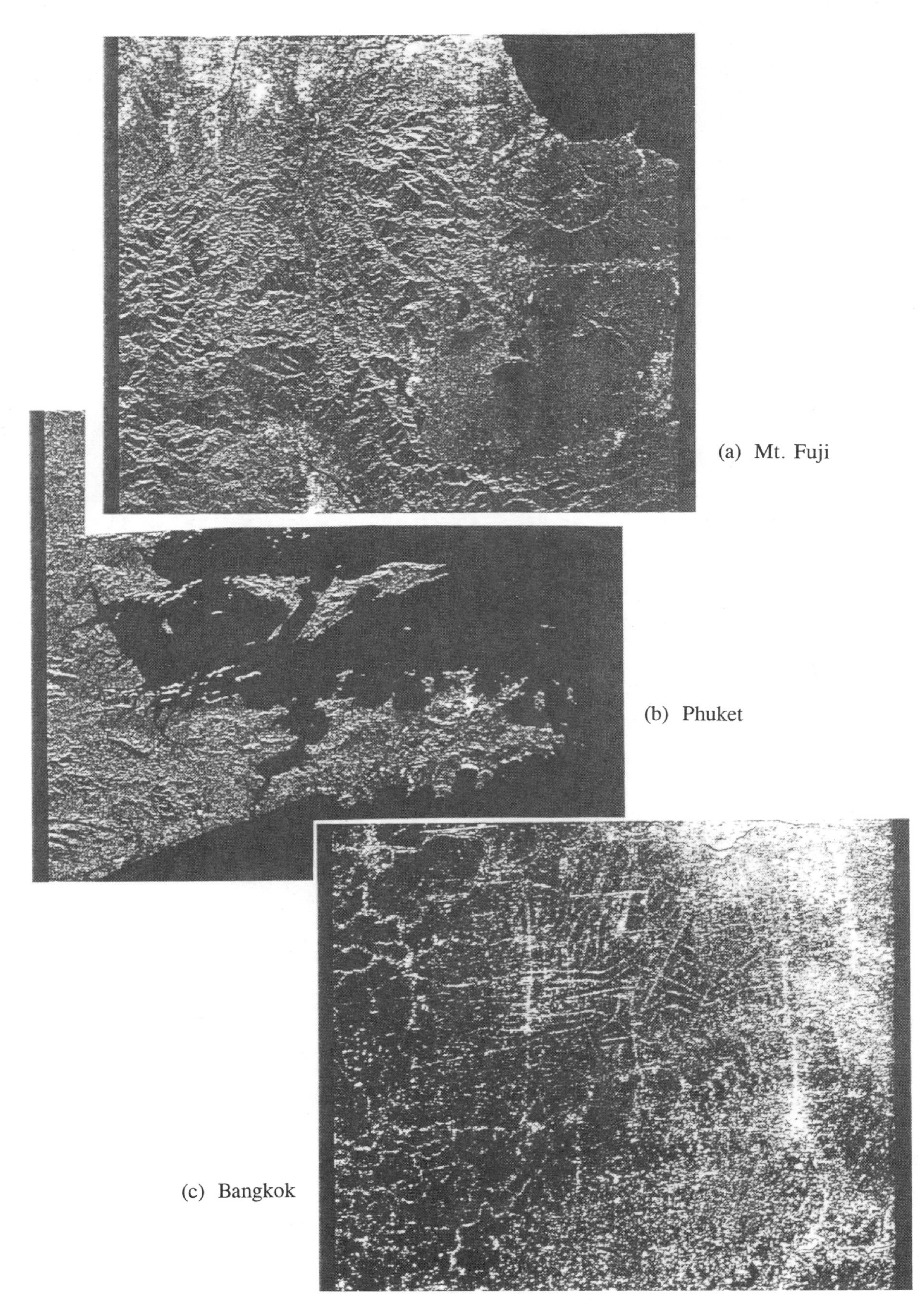

Figure IV. Examples of the output from the quick look processor

Figure V. Example of interferogram (7 July 1993 and 22 August 1993)

XIII. APPLICATION OF MICROWAVE IMAGES TO WETLAND AND COASTAL ZONE MONITORING

*Supapis Polngam**

ABSTRACT

Microwave imaging data can be a useful tool of the remote sensing community for information extraction purposes, especially for many studies in tropical areas like southern Thailand, where a heavy annual rainfall makes the acquisition of cloud-free optical images difficult. The east coast of southern Thailand has been selected as a study area for monitoring wetlands and coastal resources under the Thailand GlobeSAR project. The objective is to study how microwave images can be used to identify wetland features in the coastal zone. Data were acquired from the Canada Centre for Remote Sensing and the National Research Council of Thailand in various radar characteristic assignments. These data are simulated Radarsat, Radarsat with C-band HH polarization (C_{HH})*, ERS-1 with C-band VV polarization* (C_{VV})*, long wavelength of JERS-1 L-band HH polarization* (L_{HH}) *and optical Landsat TM. A comparison of these data indicates that uniformly high backscatter of major wetland features such as paddy field, paddy field with sugar palm and aquaculture features can be easily classified from simulated Radarsat. Difficulties arise in the case of standing tree features, including mangrove swamp forest and brackish water swamp forest. To obtain more information and further verify wetland features, multisensor simulated Radarsat and optical Landsat TM data, multisensor ERS-1, Radarsat and JERS-1 data, and multitemporal Radarsat data were applied. The results show that a mangrove swamp forest can be clearly distinguished from brackish swamp forest based on multisensor microwave and optical Landsat TM image, while multisensor microwave images of ERS-1, Radarsat and JERS-1 have diminished ability to recognize those two classes. However, JERS-1, with long wavelength* L_{HH} *is more sensitive than* C_{VV} *and* C_{HH} *to standing tree classes, which have sharp boundaries. HH polarization of C band and L band gives good response over shrimp farms and floating weeds. Wave patterns, currents and shallow sea bathymetry provide a strong radar signal from* C_{VV}*. Data fusion of multitemporal Radarsat data could reasonably support shrimp farm monitoring and rice growing stages.*

A. Introduction

Passive multispectral sensors such as Landsat and SPOT have been successfully used to map and inventory forests, land use and wetland activities in coastal resources. But there are some situations when the applicability of these sensors is severely limited, such as trying to penetrate cloud cover. It is difficult to get cloud-free data during the rainy season in tropical monsoon regions or equatorial climates, such as that where the study area is distributed. At present, radar remote sensing can provide cloud-free images and will be important for monitoring and updating forests, land use and wetlands.

The Canada Centre for Remote Sensing (CCRS) and the National Research Council of Thailand (NRCT) have cooperated under the GlobeSAR project since 1993. The purpose of this cooperation is to promote the application of Radarsat data by inviting researchers from involved organizations who wish to conduct research using the Radarsat data.

Eight study sites were selected for representing important terrain features and land activities for the purpose of learning how Radarsat data can support applications such as forest, land use,

*Thailand Remote Sensing Centre, National Research Council of Thailand, Bangkok.

agriculture, geology, geomorphology and wetland monitoring. In this paper, the application to wetland and coastal zone management is presented. A coastal resource site was selected covering an area ranging from coastal features to flood plain conditions. It is located on the east coast of southern Thailand.

At first, simulated Radarsat SAR data were supported in this research. The data were detected by the CCRS Convair-580 aircraft C-band SAR with HH polarization. The data were acquired on 5 November 1993 while field information was collected as the aircraft passed over.

A preliminary result showed clear applications in monitoring shrimp farms, paddy with sugar palm and aquaculture features. Since 1996, Radarsat data (with multimode SAR sensor including standard and fine resolution beams mode and multi-incidence angle varying 20-50°), ERS-4 data and JERS-1 data were used.

B. Project objective

The objective of the project is to study how microwave SAR data can be used as input for wetland monitoring.

C. Study area

The study area covers most of coastal southern Thailand from the west to the east, Songkhla Lake and the Thalae Luang area (see figure I). It is between, approximately, latitude 6°57'-8°12'N and longitude 100°12'-100°35'E. The physiography of the area can be described as follows: the eastern part is characterized as a coastal plain, which consists of long stretches of sandy beach, broken only by a major river, and followed by a series of parallel beach ridges. Between the beach barriers and the oldest series of beach ridges are the lagoons. They are extremely shallow (2 m in most places) and clearly undergoing rapid sedimentation; the western part is swamp. Sediments are from the numerous short rivers that no longer have direct access to the sea. As for wetland application, "wetland" is defined by the Convention on Wetlands of International Importance especially as Waterfowl Habitat, known as the Ramsar Convention, as land of marsh, fen, peatland or water, natural or artificial, permanent or temporary, with water that is static or flowing, fresh, brackish or salt, including areas of marine water, the depth of which at low tide does not exceed six metres. Figure II shows the profile of wetland land-form. The area is planted with paddy cultivation of the backswamp. Brackish swamp forest is growing in areas where slack water deposit is at lowest position, and mangrove swamp forest is located along the coastal area.

D. Data acquisition

Table 1 presents an overview of the satellite image data used.

Ground information collection was carried out during a radar pass on 5 November 1993. The data collected from field observation include various ground objects and personal interviews with local farmers.

Topographic maps at scales 1:250,000 and 1:50,000, which were reproduced by the Royal Thai Survey Department, and thematic maps of soil suitability and existing land use maps were used to support image analysis.

E. Methodology

Image analysis was done on PCI software available at the Thailand Remote Sensing Centre (TRSC), National Research Council of Thailand.

The procedure for processing of images is shown in figure III.

Table 1. Image data used in the study

Sensor and mode	Pass direction	Observation date	Band and polarization	Resolution range (m) by azimuth (m)	Swath width (km)	Incidence angle
Airborne SAR	Ascending	5 November 1993	C-band, HH, VV	6 x 6	22	0°-7°
JERS-1	Descending	7 November 1995	L-band, HH	18 x 18	75	35°
JERS-1	Descending	1 June 1997	L-band, HH	18 x 18	75	35°
ERS-1	Ascending	11 October 1993	C-band, VV	25 x 25	100 x 100	23°
ERS-1	Descending	11 November 1993	C-band, VV	25 x 25	100 x 100	23°
ERS-1	Descending	17 June 1995	C-band, VV	25 x 25	100 x 100	23°
Radarsat	Ascending	3 June 1996	C-band, HH	10 x 10	45 x 45	30°-40°
Radarsat	Ascending	7 September 1996	C-band, HH	10 x 10	45 x 45	30°-40°
Radarsat	Ascending	18 November 1996	C-band, HH	10 x 10	45 x 45	30°-40°
Landsat TM	–	5 May 1995	Optical sensor	30 x 30	185 x 185	–

In the first processing step, 16-bit SAR data were compressed to 8 bits by linear scaling technique.

Geometric correction was carried out on the original data; the simulated airborne SAR image was first rectified and then used as a base on to which both Landsat TM and SAR images were registered. Those data were registered on to a UTM grid and were resampled to 10 m x 10 m pixel size.

Next, on the simulated airborne SAR and other SAR images, speckle noise was removed using the Kuan filtering technique.

Normally, on the SAR sensor, a single black and white image is obtained, which makes it more difficult to identify an object's characteristics. So, for better utilization, SAR data fusion of satellite images (multi-sensor by combining optical Landsat TM and SAR images and multitemporal by merging Radarsat SAR images from different seasons of data acquisition) was done to enhance the data analysis of those data sets, using the best characteristics of each component (see figures IV and V).

F. Results

Based on visual interpretation of the simulated black and white image, object characteristics were identified based on the components of tone, texture, pattern and association. The tone is considered from different backscatter energy returned from each category. Table 2 illustrates a preliminary result from the simulated airborne image. But the results indicate some confusion between the standing tree classes such as brackish swamp forest, mangrove forest and mixed

Table 2. A preliminary result from the simulated airborne image

Description	Tone	Pattern/texture	Remark
1. Paddy with sugar palm	Bright	Regular pattern with coarse texture	Sugar palm trees are a dominant feature, so act as corner reflector.
2. Rain-fed paddy	Dark	Rectangular pattern, fine texture	In November, paddy is in early stage, mostly covered with water.
3. Off-season paddy	Bright	Regular pattern, fine texture	In November, off-season paddy is in fully grown stage.
4. Field preparation for rubber plantation	Dark	Regular pattern, fine texture	No return signal
5. Rubber plantation	White	Regular pattern, quite coarse texture	-
6. Floating grass	White	Regular pattern	-
7. Shrimp farm	Dark and surrounding white line	Rectangular shape	Caused by reflectance from pond ridge, appearing white around the pond.
8. Aquaculture feature	White spots	Spotted	High return signal

orchard, because of coarse surface roughness from vegetation canopies, which can provide a high return signal.

To solve this problem in separating mangrove from brackish swamp forest, data fusion of both optical Landsat TM and simulated airborne image as well as ERS-1, Radarsat and JERS-1 images, was performed by assigning the meeting of backscatter from surface roughness, while optical Landsat TM band 4 and band 5 were recognized in spectral reflectance condition for an optical and microwave multi-sensor. Satisfactory results were obtained with regard to the identification of colour and another component. Swamp forest was clearly distinguished from mangrove forest (see figure IV). Also, this data fusion image can clearly separate paddy field from paddy field with sugar palm, whereas the multi-sensor ERS-1, Radarsat and JERS-1 image has diminished ability to separate the two classes of mangrove and brackish swamp forest, for which the SAR backscatter values are nearly the same. As for digital image analysis, a supervised classification technique was performed on an experimental basis, based on a data set of multitemporal Radarsat images. Training areas were defined and the statistics of those spectral signatures were generated in mean vectors and covariance matrices. Those training sets could be tested in order to check accuracy. A higher number means higher accuracy of classification, but the number should not be less than 1.5.

The maximum likelihood classification produced 12 classes. They are water, paddy field in different stages, shrimp farm, mangrove forest, disturbed swamp forest, rubber, urban, floating grass and mixed orchard. Results were promising, indicating that mangrove forest was being replaced by shrimp farming. Paddy fields in different stages were also identified easily. However, there was some ambiguity between some types of vegetation with standing trees. Figures VI to XIV illustrate a comparison of microwave images for wetland features.

G. Conclusion

(1) Simulated Radarsat images demonstrate excellent sensitivity of radar backscatter to the surface, such as paddy with sugar palm, aquaculture features and aquatic plants;

(2) L-band JERS-1 images provide strongest radar signal with structures below the canopy, including branches and trunks; it is clearly seen in very bright tone in brackish water swamp forest;

(3) C band and L band interact strongly to return signals from trees at different height levels or in heterogeneous areas such as degraded mangrove forest;

(4) Both C_{HH} and L_{HH} give a good response over shrimp farms that were laid along the shoreline in a north-south direction, contributing to corner reflectance from dikes and the water surface as well;

(5) C_{VV} cannot detect any acquaculture features such as bamboo poles (fish traps) in rough sea, because the sea surface becomes much brighter, reducing the contrast between sea surface and those features;

(6) In both L band and C band, high backscatter of vegetation occurs over water, like mangrove forest and floating weeds, and displays sharp boundaries on images; but in C_{VV} with wind effect, sea water and those vegetation classes are differently separated;

(7) Long wavelength of L band produces high radar signal in moisture area;

(8) C_{VV} presents much information over the sea surface, including wave patterns, currents and shallow sea bathymetry;

(9) Data fusion of multi-sensor and multitemporal SAR images requires accuracy of data registration;

(10) Data fusion of multitemporal images can reasonably support vegetation monitoring;

(11) Data fusion of multipolarization can be an aid to detect vegetation types.

Bibliography

Simonett, D.S. *Manual of Remote Sensing,* 2nd ed., vol. I.

Le Toan, T., H. Iaur, E. Mougin and A. Lopes, 1998. Multitemporal and dual-polarization observations of agricultural vegetation covers by X-band SAR image, *IEEE Transactions on Geoscience and Remote Sensing,* 27(6):709-718.

Musigasarn, W. and S. Polngam, 1994. GlobeSAR input for development planning of Songkhla Lake basin. In *Proceedings* of GlobeSAR South-East Asian Regional Workshop, Bangkok, Thailand.

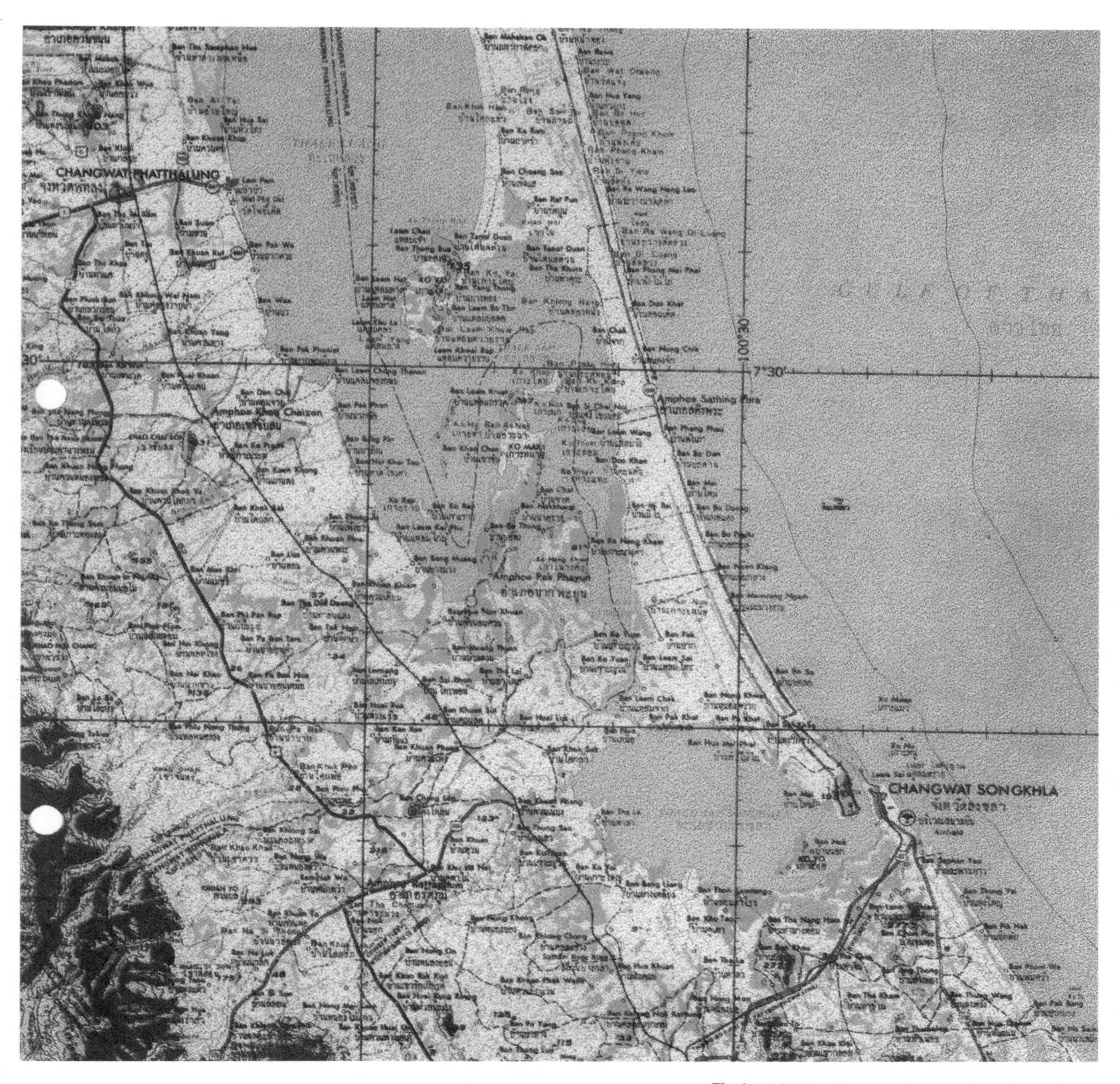

The boundaries and names shown and the designations used on this map do not imply official endorsement or acceptance by the United Nations.

Figure I. Coastal study site in Thailand

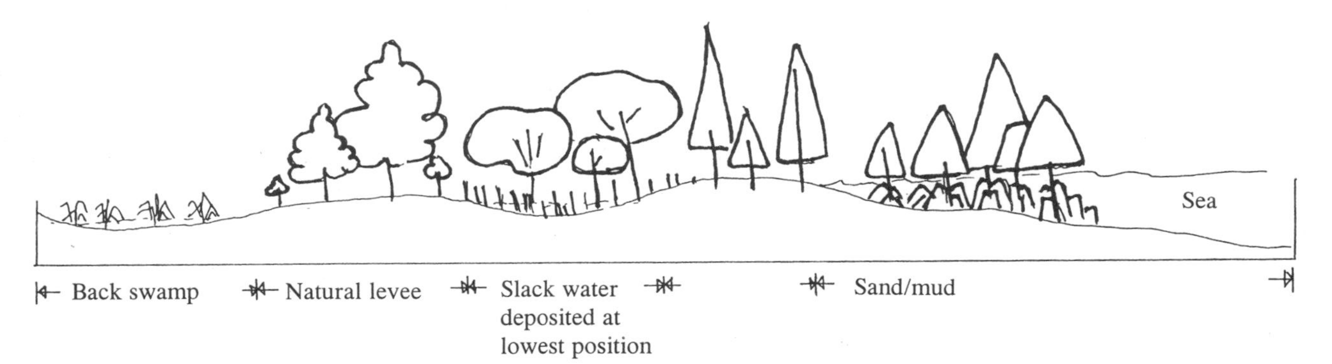

Figure II. The profile showing land-form of wetland and coastal zone

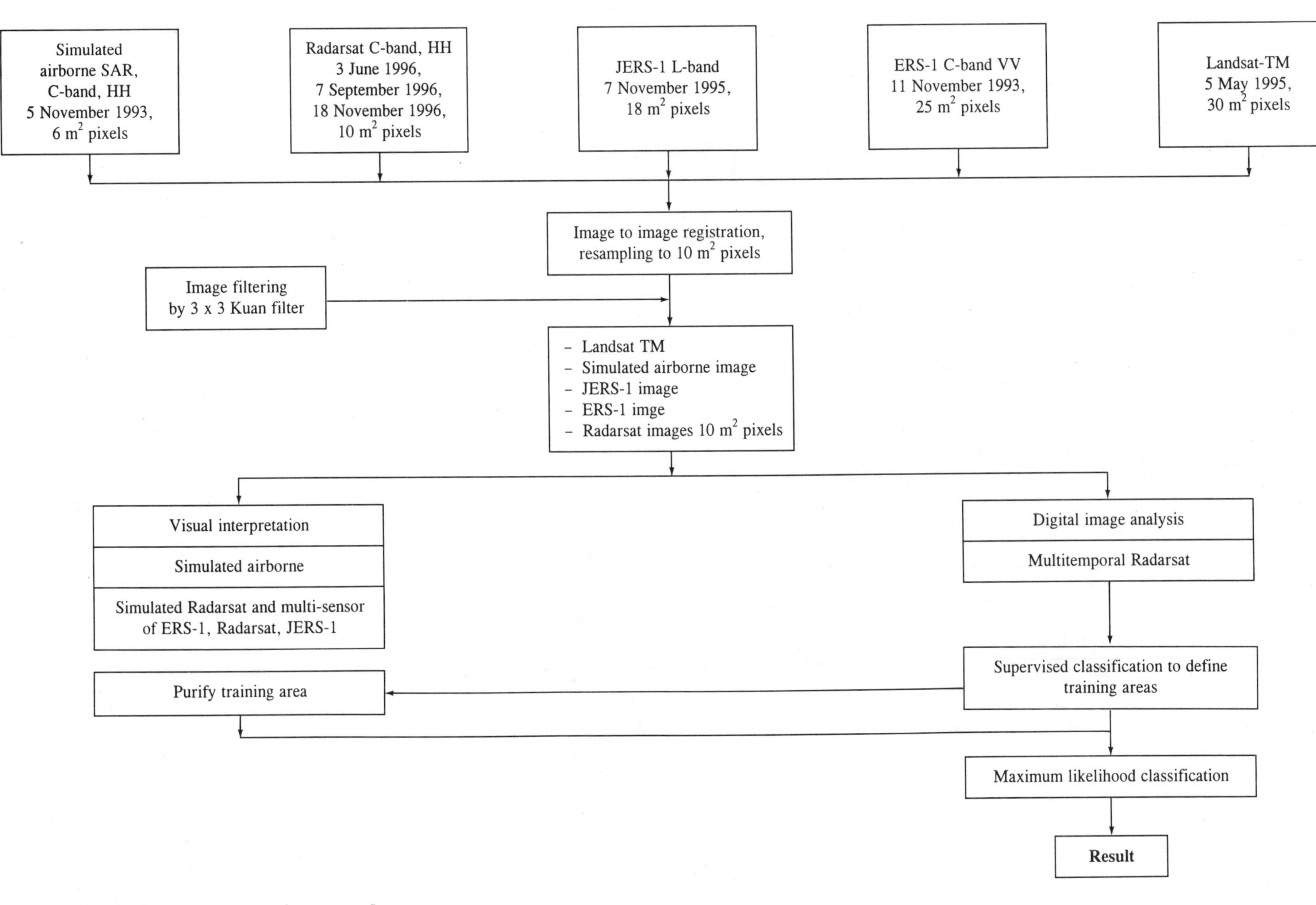

Figure III. SAR image processing procedure

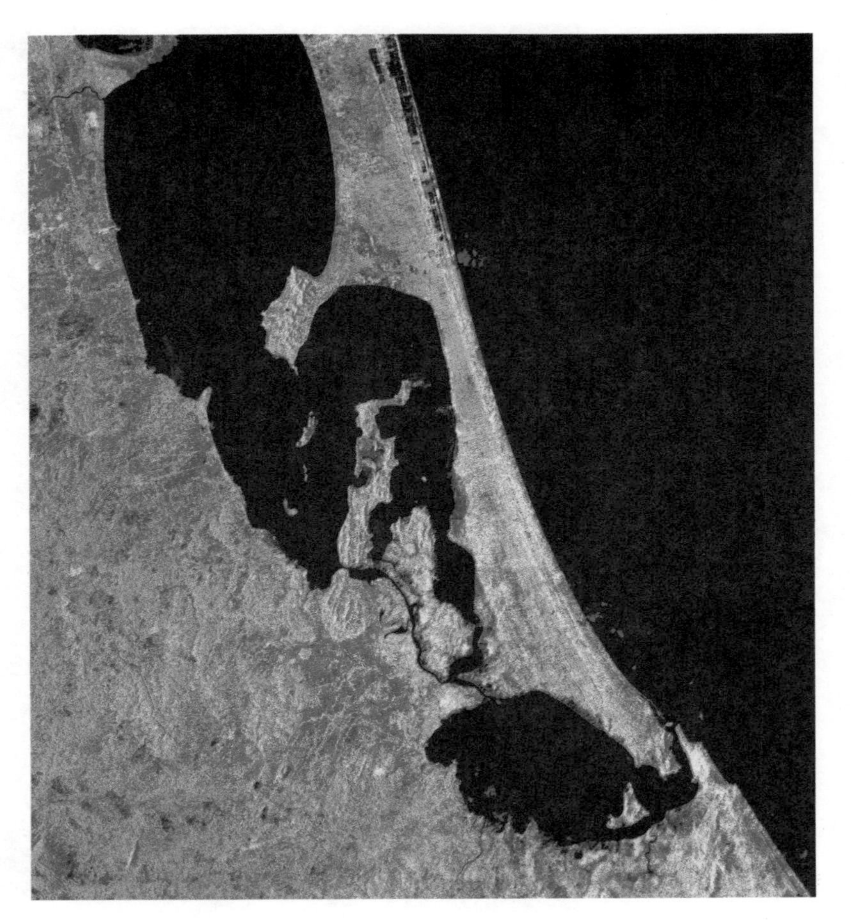

Figure IV.
Data fusion of multi-sensor, optical Landsat TM and simulated airborne image

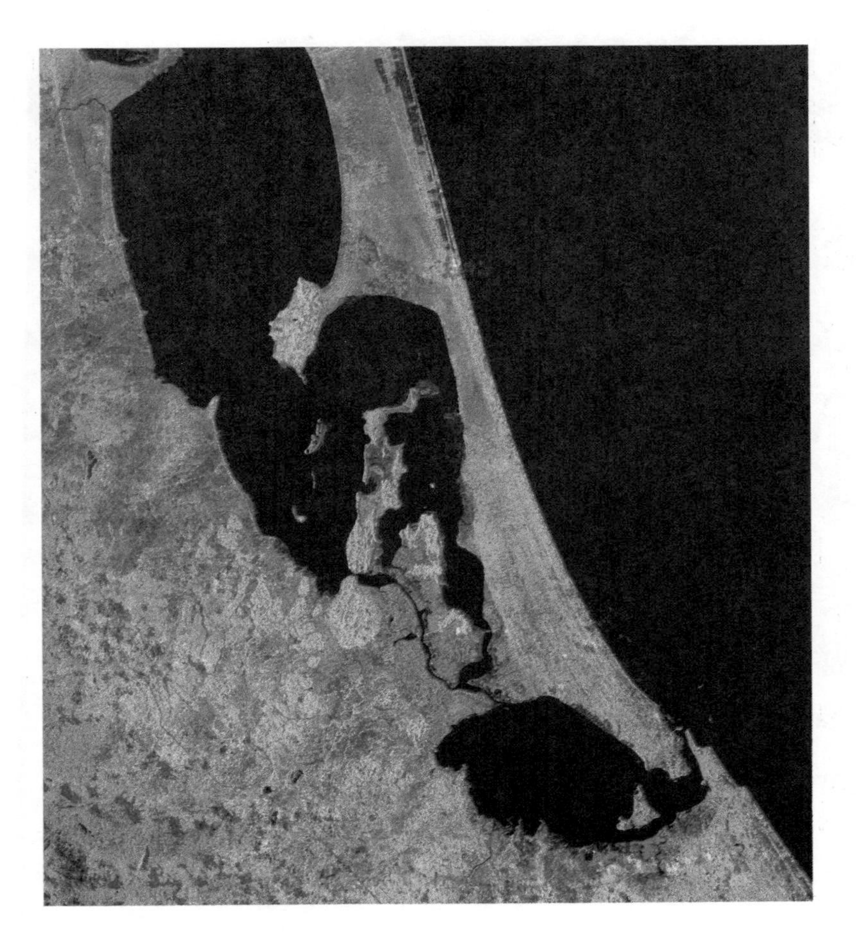

Figure V.
Data fusion of multitemporal Radarsat image

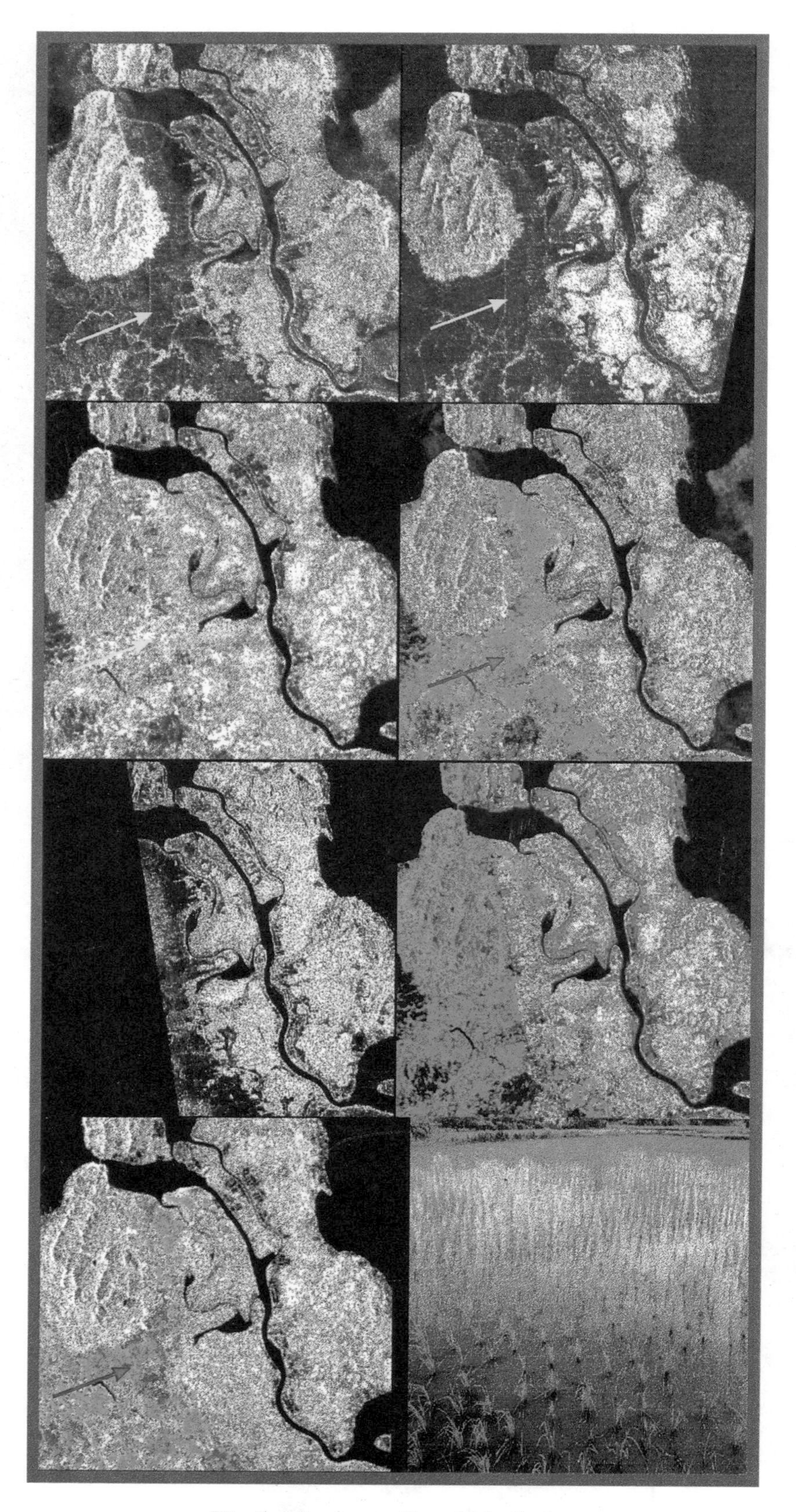

ERS-1
C_{VV}
11 November
1993

JERS-1
L_{HH}
7 November
1995

Radarsat
C_{HH}
18 November
1996

Multi-
sensor
ERS/RS/
JERS

Airborne
C_{HH}
5 November
1993

Multi-
sensor
airborne/
RS/JERS 1

Multi-
temporal
Radarsat
3 June 1996
7 September
1996
18 November
1996

Paddy field
in early stage
5 November
1993

Thalae Luang - Songkhla Lake

Figure VI. Comparison of microwave images for paddy field in early stages

ERS-1
C_{VV}
11 November
1993

JERS-1
L_{HH}
7 November
1995

Radarsat
C_{HH}
18 November
1996

Multi-
sensor
ERS/RS/
JERS

Airborne
C_{HH}
5 November
1993

Multi-
sensor
airborne/
RS/JERS 1

Multi-
temporal
Radarsat
3 June 1996
7 September
1996
18 November
1996

Paddy field
with sugar
palm

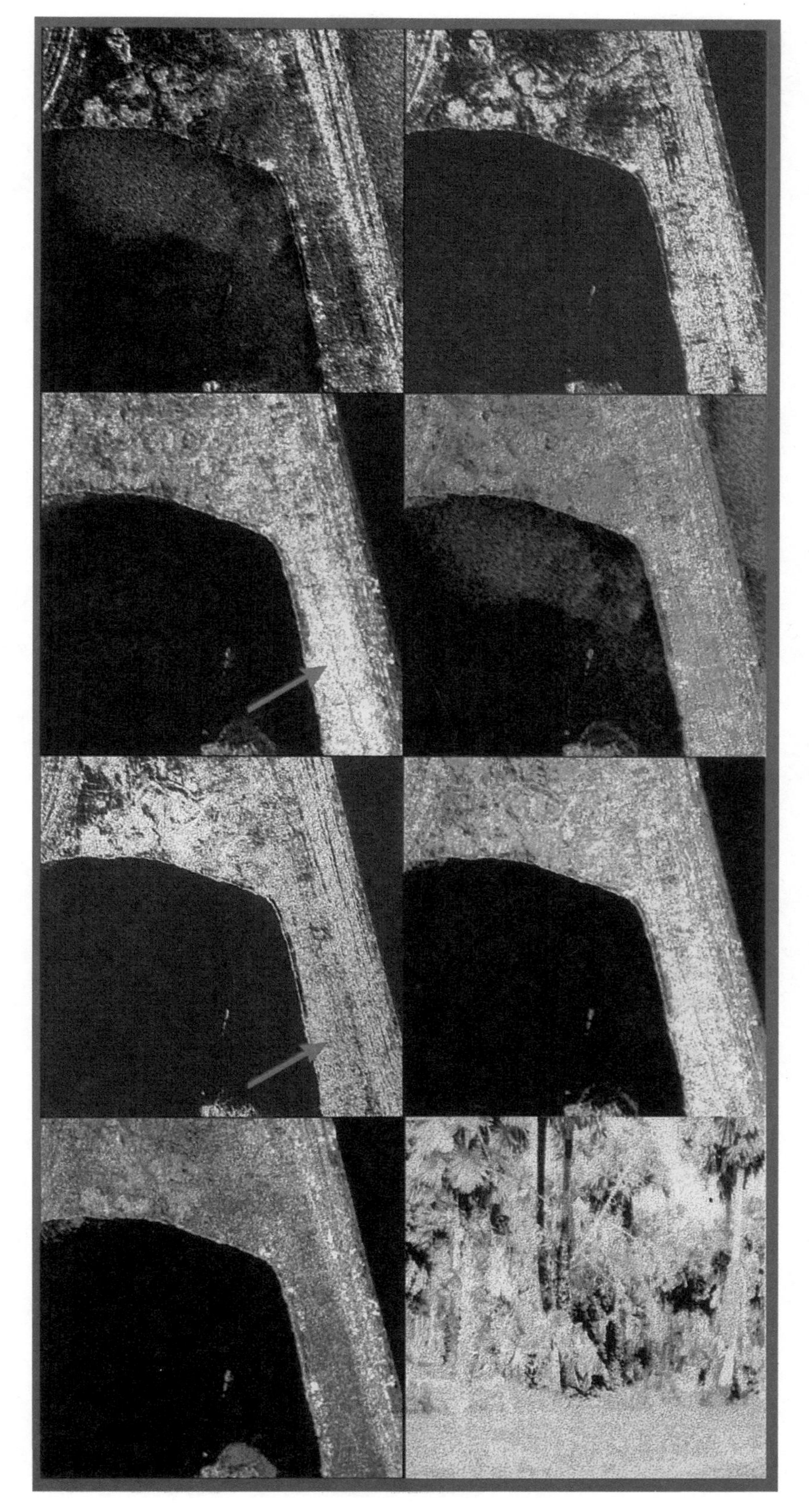

Thalae Luang - Songkhla Lake

Figure VII. Comparison of microwave images for paddy field with sugar palm

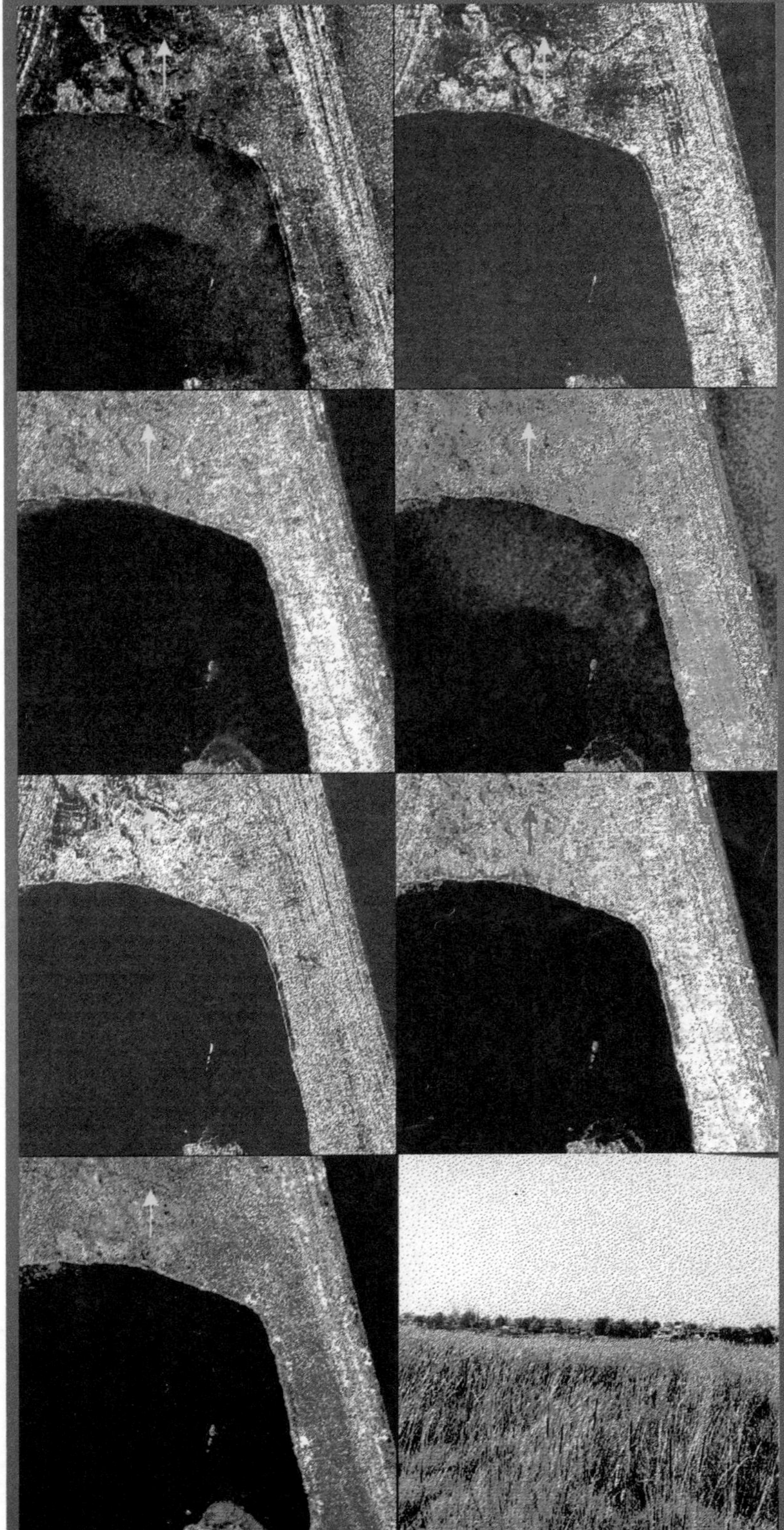

Thalae Luang – Songkhla Lake

Figure VIII. Comparison of microwave images for weed with long leaves

ERS-1
C_{VV}
11 November
1993

JERS-1
L_{HH}
7 November
1995

Radarsat
C_{HH}
18 November
1996

Multi-
sensor
ERS/RS/
JERS

Airborne
C_{HH}
5 November
1993

Multi-
sensor
airborne/
RS/JERS 1

Multi-
temporal
Radarsat
3 June 1996
7 September
1996
18 November
1996

Swamp
forest
5 November
1993

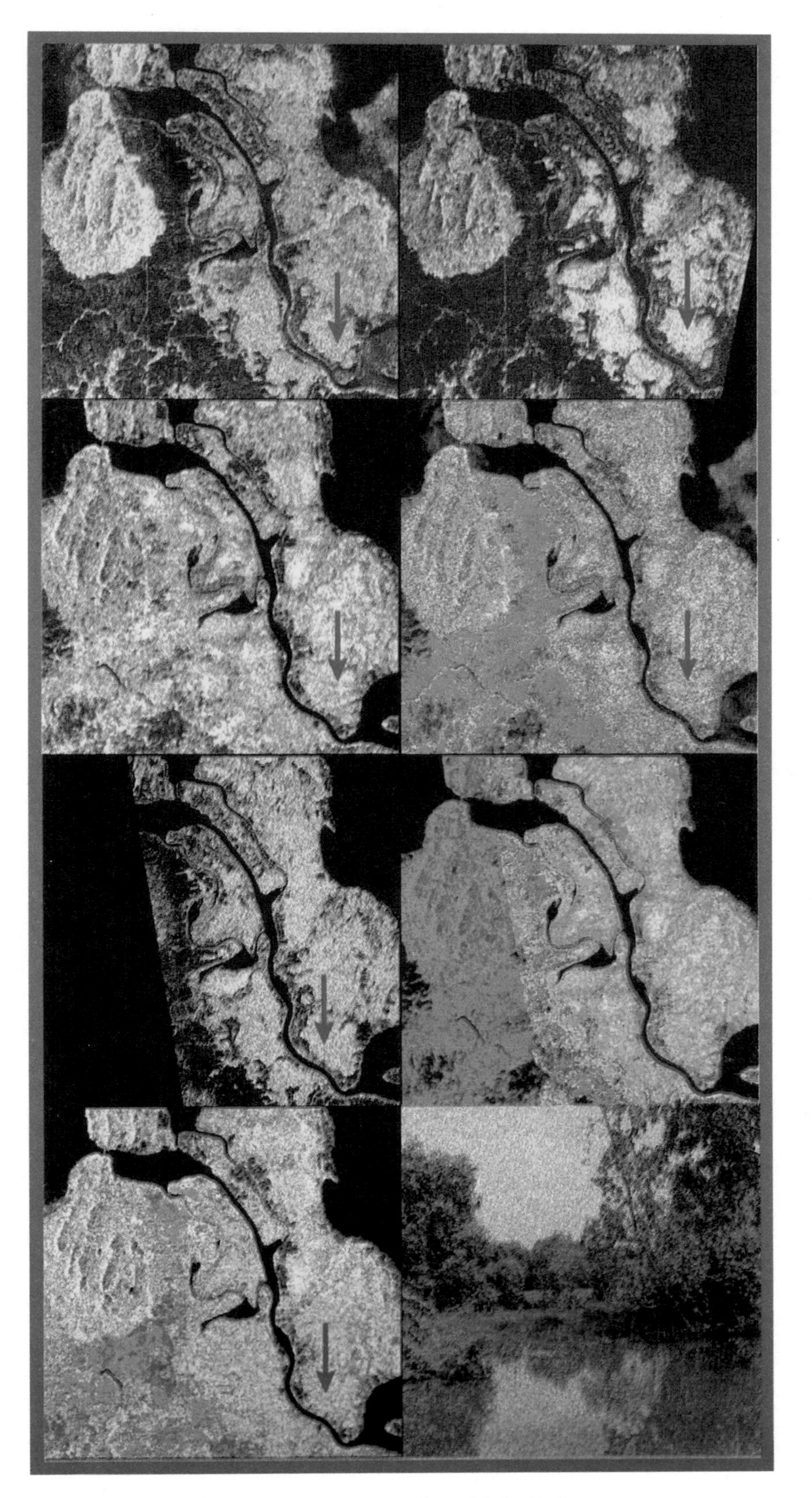

Thalae Luang – Songkhla Lake

Figure IX. Comparison of microwave images for brackish water swamp forest

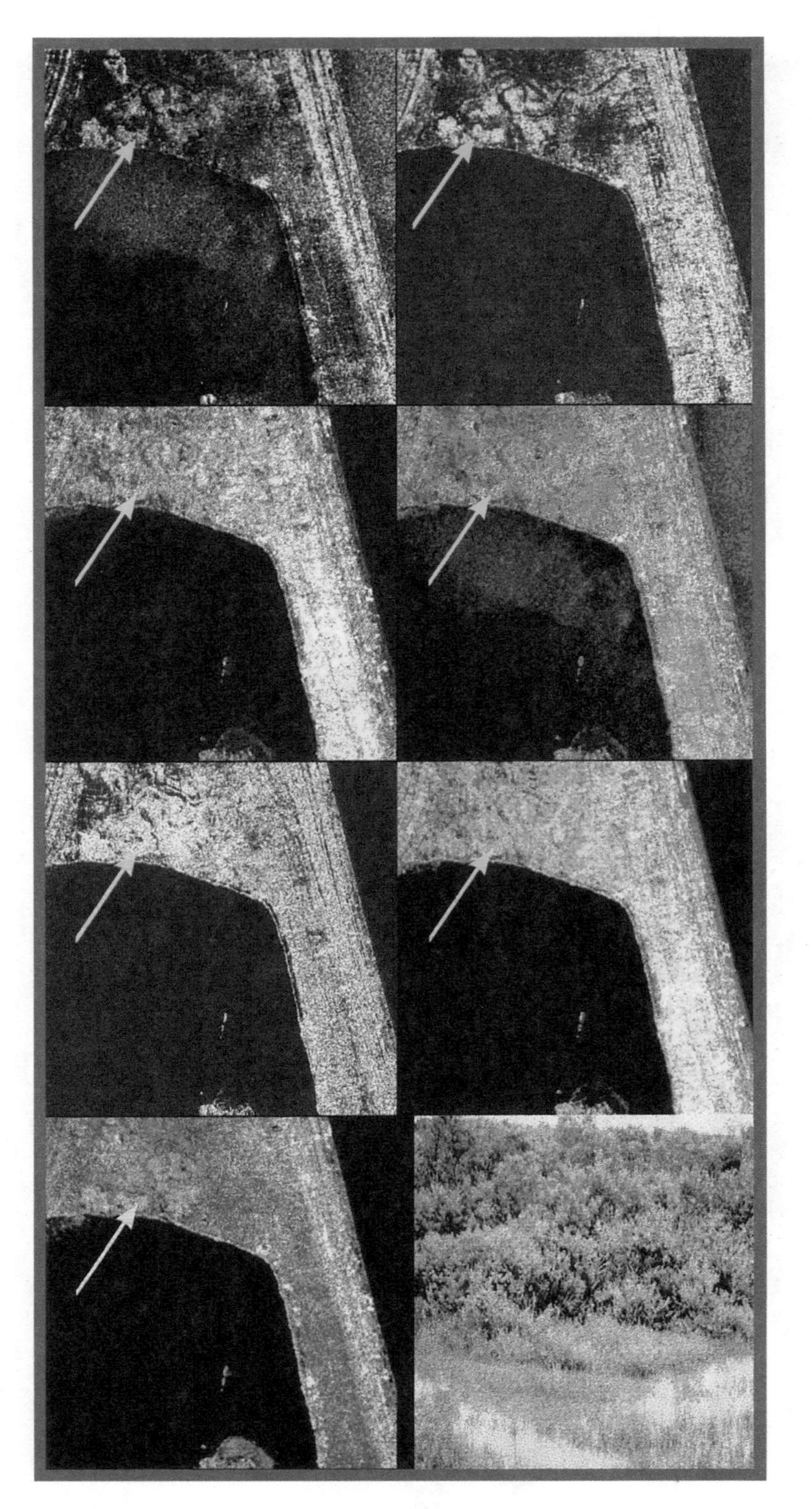

ERS-1
C_{VV}
11 November
1993

JERS-1
L_{HH}
7 November
1995

Radarsat
C_{HH}
18 November
1996

Multi-
sensor
ERS/RS/
JERS

Airborne
C_{HH}
5 November
1993

Multi-
sensor
airborne/
RS/JERS 1

Multi-
temporal
Radarsat
3 June 1996
7 September
1996
18 November
1996

Degraded
mangrove
forest
5 November
1993

Thalae Luang - Songkhla Lake

Figure X. Comparison of microwave images for degraded mangrove forest

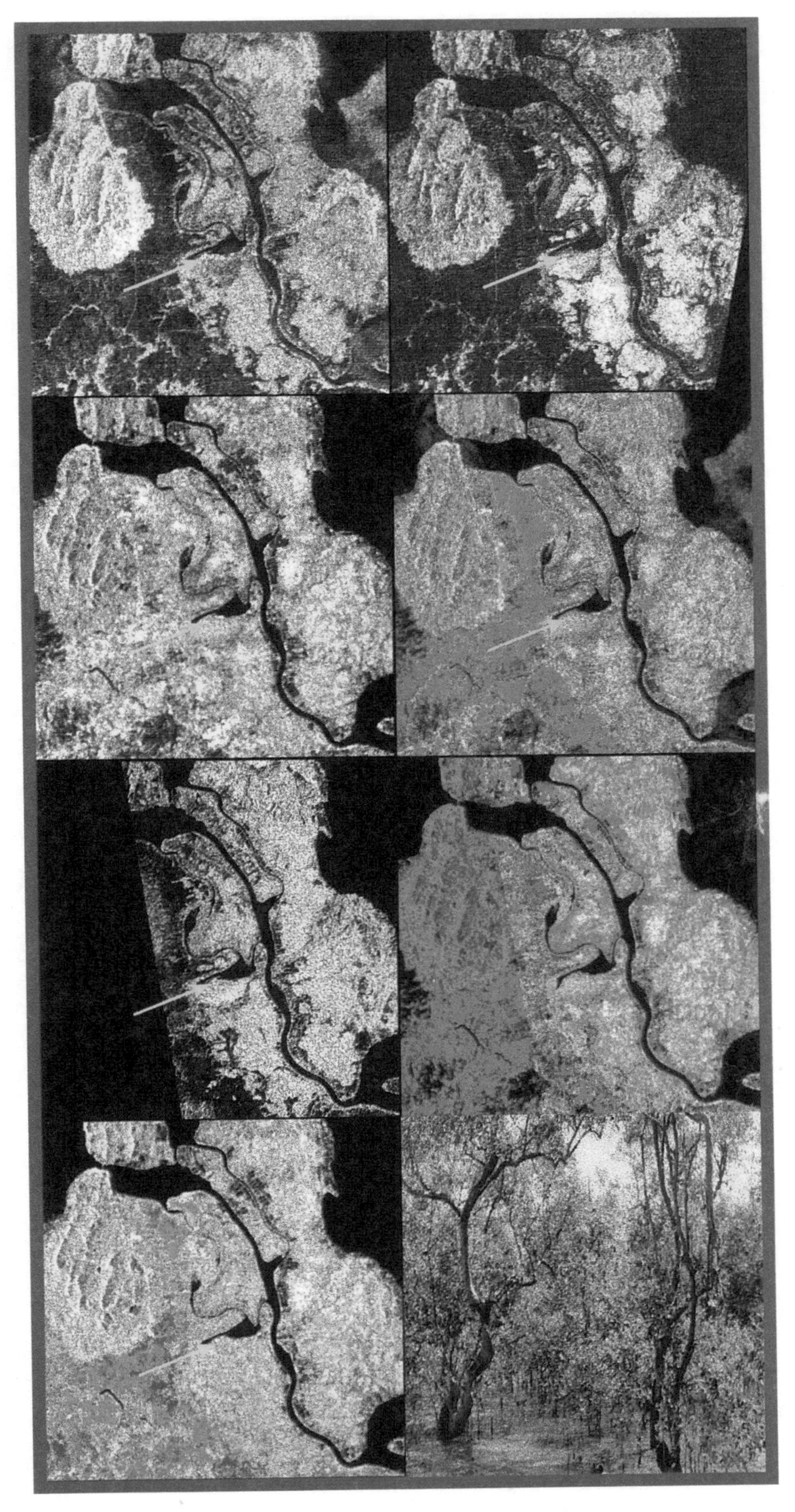

ERS-1
C_{VV}
11 November
1993

JERS-1
L_{HH}
7 November
1995

Radarsat
C_{HH}
18 November
1996

Multi-
sensor
ERS/RS/
JERS

Airborne
C_{HH}
5 November
1993

Multi-
sensor
airborne/
RS/JERS 1

Multi-
temporal
Radarsat
3 June 1996
7 September
1996
18 November
1996

Mangrove
forest
5 November
1993

Thalae Luang - Songkhla Lake

Figure XI. Comparison of microwave images for mangrove swamp forest

ERS-1
C_{VV}
11 November
1993

JERS-1
L_{HH}
7 November
1995

Radarsat
C_{HH}
18 November
1996

Multi-
sensor
ERS/RS/
JERS

Airborne
C_{HH}
5 November
1993

Multi-
sensor
airborne/
RS/JERS 1

Multi-
temporal
Radarsat
3 June 1996
7 September
1996
18 November
1996

Shrimp farm
5 November
1993

Thalae Luang – Songkhla Lake

Figure XII. Comparison of microwave images for shrimp farm

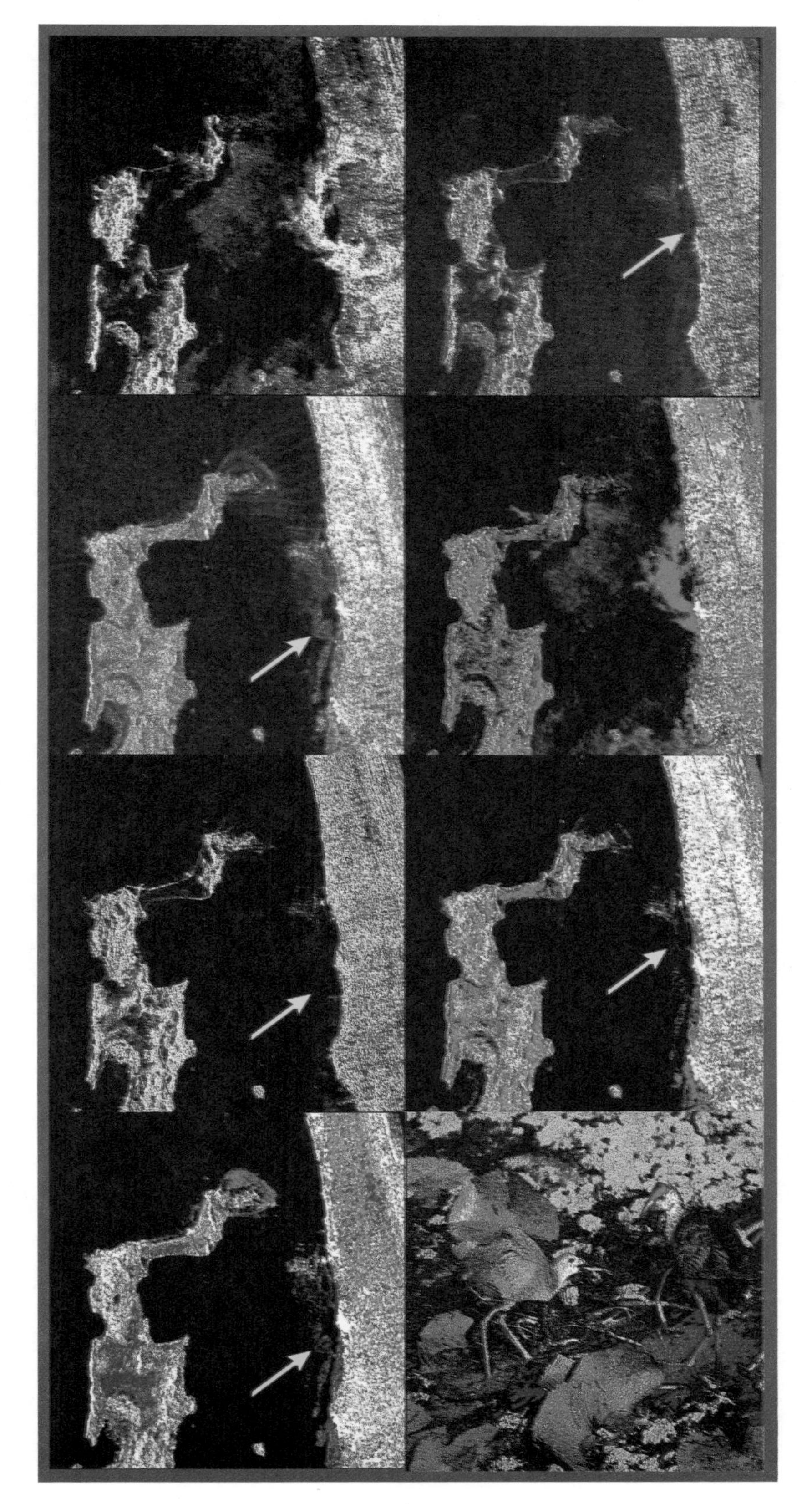

ERS-1
C_{VV}
11 November
1993

JERS-1
L_{HH}
7 November
1995

Radarsat
C_{HH}
18 November
1996

Multi-
sensor
ERS/RS/
JERS

Airborne
C_{HH}
5 November
1993

Multi-
sensor
airborne/
RS/JERS 1

Multi-
temporal
Radarsat
3 June 1996
7 September
1996
18 November
1996

Floating weed
5 November
1993

Thalae Luang – Songkhla Lake

Figure XIII. Comparison of microwave images for floating weed

ERS-1
C_{VV}
11 November
1993

JERS-1
L_{HH}
7 November
1995

Radarsat
C_{HH}
18 November
1996

Multi-
sensor
ERS/RS/
JERS

Airborne
C_{HH}
5 November
1993

Multi-
sensor
airborne/
RS/JERS 1

Multi-
temporal
Radarsat
3 June 1996
7 September
1996
18 November
1996

Fish traps
5 November
1993

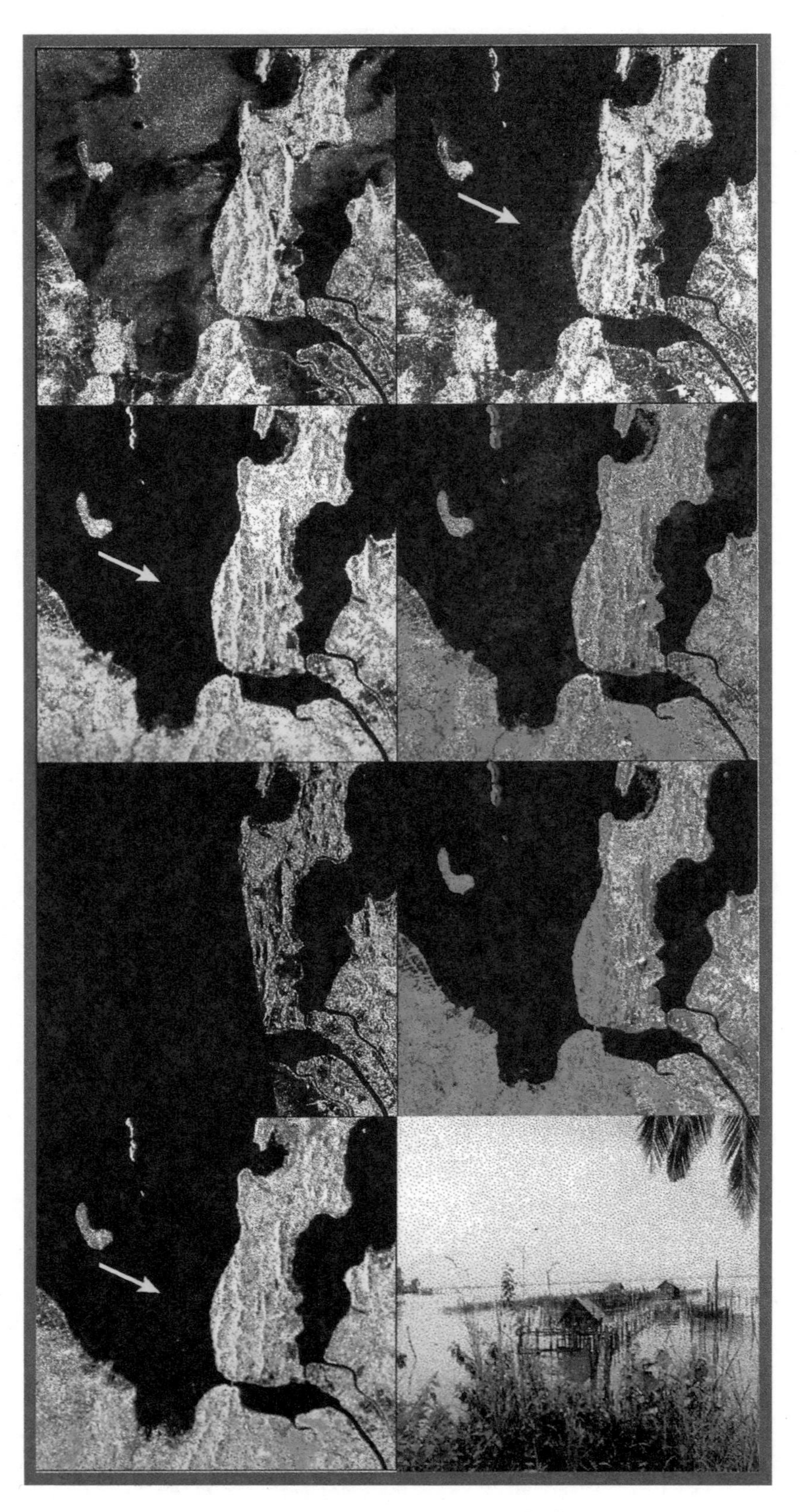

Thalae Luang – Songkhla Lake

Figure XIV. Comparison of microwave images for aquaculture feature/fish traps

ANNEX

LIST OF PARTICIPANTS

Bangladesh

Mr Nazmul Hoque, Chief Scientific Officer, Space Research and Remote Sensing Organization (SPARRSO), Mohakash Biggyan Bhaban, Agargaon, Dhaka 1207 [Tel: 880-2-323942, Fax: 880-2-813080, Telex: 642215 SRS BJ, E-mail: sparrso@bangla.net]

Cambodia

Mr Mao Sothun, Remote Sensing and Mapping Team leader, Integrated Resource Information Centre (IRIC), Ministry of Public Works, 200 Norodom Blvd., Phnom Penh [Tel: 725-107, 725-007, Mob: 015-914 336]

China

Mr Wu Yongshun, Assistant Consultant, Department of Spatial Planning and Regional Economy, State Planning Commission of China, 38 S. Yuetan Street, Beijing 100824 [Tel: 10-68502740, Fax: 10-68511589, 68502747, E-mail: wuys@sicdb.cei.go.cn, dsprezhc@mx.cei.go.cn]

Germany

Professor Gottfried Konecny, Professor for Photogrammetry, Remote Sensing and Spatial Information Systems, Institute for Photogrammetry and Engineering Surveys, University of Hannover, Nienburger Strasse 1, D-30167 Hannover [Tel: 49-511-762-2481, Fax: 49-511-762-2483, E-mail: gko@ipi.uni. hannover.de]

Ms Lieselatte Konecny, Nienburger Strasse 1, D-30167 Hannover [Tel: 49-511-762-2481, Fax: 49-511-762-2483]

Japan

Mr Sohsuke Gotoh, Special Assistant to the President, Earth Observation Planning Department, NASDA, World Trade Centre Building 27 Fl., 2-4-1 Hamamatsu-cho, Minato-ku, Tokyo, 105-60 [Tel: 81-3-3438-6330, Fax: 81-3-5401-8702]

Mr Takashi Moriyama, Senior Engineer, Earth Observation Planning Department, NASDA – Japan, World Trade Centre Building 27 Fl., 2-4-1 Hamamatsu-cho, Minato-ku, Tokyo, 105-60 [Tel: 81-3-3438-6332, Fax: 81-3-5401-8702]

Ms Yuko Nagata, Staff, Earth Observation Planning Department, NASDA World Trade Centre Building 27 Fl., 2-4-1 Hamamatsu-cho, Minato-ku, Tokyo, 105-60 [Tel: 81-3-3438-6330, Fax: 81-3-5401-8702]

Mr Toshiaki Hashimoto, Associate Senior Engineer, Earth Observation Research Centre, NASDA, Roppongi First Building 13 Fl., 1-9-9 Roppongi, Minatoku, Tokyo 106 [Tel: 81-3-3224-7072, Fax: 81-3-3224-7052]

Mr Masaru Tsukamoto, Director, Bangkok Office, NASDA of Japan, Special Counsellor to the Japanese Minister of State for Science and Technology, B.B. Building 13 Fl., Room 12, 54 Asoke Road, Sukhumvit 21, Bangkok 10110, Thailand [Tel: 66-2-260-7026, Fax: 66-2-260-7027, E-mail: nasdatha @ksc15.th.com]

Mr Kohei Cho, Assistant Professor, Research and Information Centre, Tokai University, 2-28-4 Tomigaya, Shibuyaku, Tokyo, 151 [Tel: 81-3-3481-0611, Fax: 81-3-3481-0610]

Mr Shiro Ochi, Assistant, Institution of Industrial Science, University of Tokyo, 7-22-1 Roppongi, Minatoku, Tokyo, 106 [Tel: 81-3-3402-6231, Fax: 81-3-3479-2762]

Mr Makoto Ono, Senior Research Scientist, Promotion Department, RESTEC, Roppongi First Building 12 Fl., 1-9-9 Roppongi, Minato-ku, Tokyo 106 [Tel: 81-3-5561-8778, Fax: 81-3-5561-9572]

Mr Shoji Takeuchi, Senior Researcher, Research Department, RESTEC, Roppongi First Building 8 Fl., 1-9-9 Roppongi, Minato-ku, Tokyo 106 [Tel: 81-3-5561-8761, Fax: 81-3-5561-9541]

Ms Setsuko Negishi, Manager, Training and Dissemination, Promotion Department, RESTEC, Roppongi First Building 12 Fl., 1-9-9 Roppongi, Minato-ku, Tokyo 106 [Tel: 81-3-5561-9776, Fax: 81-3-5561-9541]

Ms Chikako Iguchi, Staff, Training and Dissemination, Promotion Department, RESTEC, Roppongi First Building 12 Fl., 1-9-9 Roppongi, Minato-ku, Tokyo 106 [Tel: 81-3-5561-9776, Fax: 81-3-5561-9541]

Mr Kazuhiko Terao, Administrative Assistant, United Nations Centre for Regional Development (UNCRD), Nagoya International Center Building 6 Fl., 1-47-1 Furu-machi, Nakamuraku-gun, Nagoya [Tel: 052-561-9375, Fax: 052-561-9377]

Mr Atsushi Ono, Exhibition - RESTEC

Ms Makiko Yoshida, Exhibition - RESTEC

Mr Toshimasa Naishinaga, Exhibition - Assistant Co.

Mr Shuji Yamashita, Exhibition - NEC

Mr Ben Eguchi Tsutonue, President Scitek Incorporated (Exhibition - SciTek Bangkok), Kamiyacho Square Building 7 Fl., 1-7-3 Azabudai, Minato-ku, Tokyo 106 [Tel: 81-3-5572-7491, Fax: 81-3-5572-7490, E-mail: ben@sti.co.jp]

Mr Vu Duc Huan, Exhibition - SciTek, Bangkok

Lao People's Democratic Republic

Mr Vayaphat Thattamanivong, Head of Cartography and Mapping, Soil Survey and Land Classification Centre, Department of Agriculture and Extension, Ministry of Agriculture and Forestry, Dongdok Road, Vientiane [Tel: 856-21-732047, Fax: 856-21-732047]

Malaysia

Mr Sahibi Mokhtar, Assistant Director, Soil Management Division, Department of Agriculture, Jalan Sultan Salahuddin, 50632, Kuala Lumpur [Tel: 4403531, Fax: 2947336, E-mail: doa30@pop.moa.my]

Myanmar

Mr U Ohn Gyaw, Deputy Director, Department of Meteorology and Hydrology, Kaba-Aye Pagoda Road, Mayanoon P.O. 11061, Yangon [Tel: 675-3011665, Fax: 675-3011691]

Philippines

Mr Eriberto C. Argete, Director III, Policy Studies Division, Department of Environment and Natural Resources, Visayas Avenue, Diliman, Quezon City, Philippines

Sri Lanka

Mr H. Manthrithilake, Director, Environment and Forest Conservation Division, Mahaweli Authority of Sri Lanka, Dam Site, Polgolla, Sri Lanka [Tel: 08-499275, 08-499728, Fax: 08-234950, 499727, E-mail: efodmasl@slt.lk]

Thailand

Ms Darasri Dowreang, Chief, Application Group, Remote Sensing Division, NRCT, 196 Phahonyothin Road, Chatuchak, Bangkok 10900 [Tel: 66-2-562-0428, Fax: 66-2-561-3035]

Ms Supapis Polngam, Research Scientist, Thailand Remote Sensing Centre, NRCT, 196 Phahonyothin Road, Chatuchak, Bangkok 10900 [Tel: 66-2-579-0345, Fax: 66-2-579-5618]

Ms Sirin Kawla-ierd, The Office of HRH, Princess Maha Chakri Sirindhorn's Projects, Bureau of the Royal Household, Chitralada Palace, Dusit, Bangkok 10303 [Tel: 66-2-281-3921, Fax: 66-2-579-3923]

Mr Vithya Srimanobhas, Senior Lecturer and Assistant Dean, Mahidol University, Faculty of Environment and Resource Studies, Salaya, Nakorn Pathom 73170 [Tel: 66-2-441-0211-6, Fax: 66-2-441-9509-10, E-mail: envsm@mucc.ac.th]

Viet Nam

Professor Nguyen Van Hieu, President of National Centre for Science and Technology (NCST), Nghia Do, Tu Liem, Hanoi [Tel: 84-4-8361779, Fax: 84-4-8352483]

Professor Tran Manh Tuan, Deputy Director General, NCST, Nghia Do, Tu Liem, Hanoi [Tel: 84-4-8361780, Fax: 84-4-8352483, E-mail: tuan@nghiado.ac.vn]

Mr Trinh Quang, Director, International Cooperation Department, NCST, Nghia Do, Tu Liem, Hanoi [Tel: 84-4-8355607, Fax: 84-4-8352483, E-mail: tquang@netnam.org.vn]

Mr Chu Tri Thang, International Cooperation Department, NCST, Nghia Do, Tu Liem, Hanoi [Tel: 84-4-8355607, Fax: 84-4-8352483]

Mr Pham Van Cu, Deputy Director of VTGEO, Institute of Geology, NCST, Nghia Do, Tu Liem, Hanoi [Tel: 84-4-8351493, Fax: 84-4-8250000, E-mail: cu@rg-igl.ac.vn]

Mr Nguyen Manh Cuong, Head of Remote Sensing Section, Forest Resources and Environment Centre, Forestry Institute of Planning and Investigation (FIPI), Ministry of Agriculture and Rural Development (MARD), FIPI, Van Dien, Thanh Tri, Hanoi [Tel: 84-4-8615513, 8613858, 8615388, Fax: 84-4-8612881]

Ms Le Thi Nhu Tam, International Relation Office, NCST, 1 Mac Dinh Chi, Q.1, Ho Chi Minh City [Tel: 84-8-8243291, Fax: 84-8-8222068]

Mr Le Ngoc Xuyen, Chief, Southern Branch, NCST, 1 Mac Dinh Chi, Q.1, Ho Chi Minh City [Tel: 84-8-8243291, Fax: 84-8-8222068]

Mr Nguyen Viet Chien, Head, Remote Sensing Division, Sub-Institute of Physics, NCST, 175 Hai Ba Trung, Ho Chi Minh City [Tel: 84-8-8234915, Fax: 84-8-8234133, E-mail: Vientham@netnam2.Org.Vn]

Mr Nguyen Dinh Duong, Department of Remote Sensing Technology and GIS, Institute of Geography, NCST, Nghia Do, Tu Liem, Hanoi [Tel: 84-4-8358333 ext. 1221, Fax: 84-4-8352483, E-mail: duong@igg.ac.vn]

Ms Tran Minh Y, Head, Department of Remote Sensing Technology and GIS, Institute of Geography, NCST, Nghia Do, Tu Liem, Hanoi [Tel: 84-4-8358333 ext. 1221, Fax: 84-4-8352483, E-mail: my@igg.ac.vn]

Mr Lam Dao Nguyen, Remote Sensing Division, Sub-Institute of Physics, NCST, 175 Hai Ba Trung, Ho Chi Minh City, [Tel: 84-8-8234915, Fax: 84-8-8234133]

Ms Truong Thi Hoa Binh, Department of Remote Sensing Technology and GIS, Institute of Geography, NCST, Nghia Do, Tu Liem, Hanoi [Tel: 84-4-8358333 ext. 1221, Fax: 84-4-8352483]

Ms Le Kim Thoa, Department of Remote Sensing Technology and GIS, Institute of Geography, NCST, Nghia Do, Tu Liem, Hanoi [Tel: 84-4-8358333 ext. 1221, Fax: 84-4-8352483, E-mail: thoa@igg.ac.vn]

Mr Tran Duc Thanh, Head, Marine Geo-environment Department, Haiphong Sub-Institute of Oceanography, NCST, 246 Danang St., Haiphong City [Tel: 84-31-846523, Fax: 84-31-846521, E-mail: tdthanh @hio.ac.vn]

Mr Dinh Van Huy, Head of Remote Sensing and GIS Group, Haiphong Sub-Institute of Oceanography, NCST, 246 Danang St., Haiphong City [Tel: 84-31-846523, Fax: 84-31-846521]

Ms Ta Thi Kim Oanh, Sub-Institute of Geography, NCST, 1 Mac Dinh Chi, Q.1, Ho Chi Minh City [Tel: 84-8-8220829, Fax: 84-8-8224895]

Ms Tran Thi Van, Sub-Institute of Geography, NCST, 1 Mac Dinh Chi, Q.1, Ho Chi Minh City [Tel: 84-8-8299618, Fax: 84-8-8224895]

Mr Nguyen Xuan Lang, Remote Sensing Division, Institute of Physics, NCST, Thu Le, Ba Dinh, Hanoi [Tel: 84-4-8347206, Fax: 84-4-8349050]

Mr Nguyen Van Vinh, Remote Sensing Division, Sub-Institute of Physics, NCST, 175 Hai Ba Trung, Ho Chi Minh City [Tel: 84-8-8234915, Fax: 84-8-8234133]

Mr Nguyen Duy Vi, Remote Sensing Division Sub-Institute of Physics, NCST, 175 Hai Ba Trung, Ho Chi Minh City [Tel: 84-8-8234915, Fax: 84-8-8234133]

Mr Le Phat Quoi, Head of Scientific Management Division, Department of Science, Technology and Environment (DOSTE) Long An Province, Ministry of Science, Technology and Environment (MOSTE), 4A National Road No. 1A, Phuong 4, Tanan Town, Long An Province [Tel: 84-72-822809, Fax: 84-72-822093]

Ms Nguyen Thi Thuy, DOSTE of Dong Thap Province, MOSTE, Vo Truong Toan St., F.1, Cao Lanh Town, Dong Thap [Tel: 84-067-851543, 853551]

Ms Nguyen Thi Anh Hong, Remote Sensing and GIS Centre, Sub-National Institute of Agriculture Planning and Projection (Sub-NIAPP), Ministry of Agriculture and Rural Development, 20 Vo Thi Sau, Q.3, Ho Chi Minh City [Tel: 84-8-8204030, Fax: 84-8-8204031]

Mr Tran Tuan Tu, Remote Sensing and GIS Centre, Geological Mapping Division No. 6, Department of Geology, Ministry of Industry, 200 Ly Chinh Thang, Q.3, Ho Chi Minh City [Tel: 84-4-8437610]

Mr Nguyen Thanh Hung, Remote Sensing and GIS Laboratory, Sub-Institute of Geography, NCST, 1 Mac Dinh Chi, Q.1, Ho Chi Minh City [Tel: 84-8-8299618, Fax: 84-8-8222068]

Mr Doan Van Tuan, Ho Chi Minh City Television

Mr Dong Quang Thinh, Ho Chi Minh City Television

Mr Nguyen Minh Hue, Ho Chi Minh City Television

AIT, Bangkok

Professor Shunji Murai, Chair Professor, Space Technology Applications and Research (STAR) Programme, AIT, P.O. Box 4, Klong Luang 12120, Pathum Thani, Thailand [Tel: 66-2-524-5599, Fax: 66-2-524-6147]

Ms Taeko Murai, AIT, P.O. Box 4, Klong Luang 12120, Pathum Thani, Thailand

Mr Robert Schumann, ESA Representative to South-East Asia, EC-ASEAN, Project Manager, AIT, P.O. Box 4, Klong Luang 12120, Pathum Thani, Thailand [Tel: 66-2-524-5598, Fax: 66-2-524-5596]

ESCAP

Mr He Changchui, Chief, Space Technology Applications Section, Environment and Natural Resources Management Division, ESCAP, United Nations Building, Rajadamnern Avenue, Bangkok 10200, Thailand [Tel: 66-2-2881234, Fax: 66-2-2881000]